TROPICAL GRASSY WEEDS

CASAFA Report Series

Report Series of CASAFA (Committee on the Application of Science to Agriculture, Forestry and Aquaculture) of the International Council of Scientific Unions (ICSU), published by CAB International:

1. Drought Resistance in Cereals
2. Tropical Grassy Weeds
3. Rapid Propagation of Fast-growing Woody Species

Tropical Grassy Weeds

Edited by

F.W.G. Baker

Scientific Secretary
CASAFA

and

P.J. Terry

Tropical Weeds Unit
AFRC Institute of Arable Crops Research
Long Ashton Research Station
Long Ashton, Bristol, UK

C·A·B INTERNATIONAL

for

CASAFA

(Committee on the Application of Science to Agriculture, Forestry and Aquaculture)

C·A·B International
Wallingford
Oxon OX10 8DE
UK

Tel: Wallingford (0491) 32111
Telex: 847964 (COMAGG G)
Telecom Gold/Dialcom: 84: CAU001
Fax: (0491) 33508

British Library Cataloguing in Publication Data
Tropical grassy weeds.
1. Tropical regions. Weeds
I. Baker, F.W.G. II. Committee on the Application of Science to Agriculture, Forestry and Aquaculture III. Series
632.580913

ISBN 0-85198-709-5

Typeset by Leaper & Gard Limited, Bristol
Printed and bound in Great Britain by
Redwood Press Limited, Melksham, Wiltshire

Contents

Preface

F.W.G. Baker
Scientific Secretary, CASAFA

This volume, the second in the CASAFA Report Series, is based on the proceedings of the CASAFA Workshop on Integrated Control of Grassy Weeds in Tropical Subsistence Farming Systems held at the International Centre for Insect Physiology and Ecology in Nairobi from 5th to 8th February 1990. The subject was chosen because weeds cause larger losses in crop production than any other factor. Grassy weeds in particular present a problem because of their similarity to cereal crop plants. In some cases, for example with *Oryza rufipogon* and *Sorghum halepense*, the weeds cause problems because they belong to the same biological species as the crop and have similar biochemical, physiological and morphological characteristics. These similarities not only create difficulties in producing an effective chemical control agent for the weed but also unwanted and undesirable characteristics can be introduced into the crop through hybridization. In addition the similarity between the crop and the weed can create problems of identification when hand weeding.

The use of similar management practices for the same crop over several years creates a selection pressure on the weeds. This may result in the development of weeds that are better adapted and more tolerant of these practices. This is a particular problem for small farmers in the Third World where management tends to be traditional and therefore remains the same over many years.

CASAFA has adopted a policy to make more than 100 free copies of each volume in the CASAFA Report Series available to institutions in developing countries so that the results and recommendations of the workshops may be more readily applied in the Third World, resulting in increased harvests. This may well be of particular importance if climate change alters the environmental stresses on crop production systems.

The Members of CASAFA are grateful to the Director of the International Centre for Insect Physiology and Ecology, the African Academy of Science, the Special Programmes Section of the Netherlands Ministry for Development Cooperation, the International Rice Research Institute, the Australian CSIRO and the International Union of Pure and Applied Chemistry for their interest, support and contributions to the Workshop. The participants would like to express their appreciation to Steve C. Collins of Monsanto for organizing a field trip to plantations and small farms in the Nairobi region which provided an opportunity to see some of the problems of grassy weeds under East African conditions.

Introduction

J. Dekker

Chairman, Programme Committee

CASAFA, the Committee on the Application of Science to Agriculture, Forestry and Aquaculture, was established in 1979 by the International Council of Scientific Unions (ICSU), the largest non-governmental scientific body in the world. CASAFA's purpose is to identify areas of research in which scientists from developed and developing countries can co-operate. It provides co-ordinating mechanisms to focus attention on and address the critical research problems of developing countries. Among other activities it has committed itself to the pursuit of the following Brundtland Commission Recommendations:

1. promotion of sustainable agriculture:
2. more efficient use of dryland areas, specifically in relation to drought and saline tolerance;
3. biological alternatives to agricultural chemicals, especially pesticides.

One of the activities of CASAFA is the organization of symposia and workshops, where pure and applied scientists come together to look for practical and novel solutions to problems. For example, workshops were held on Sustainable Agriculture in Rice Farming in May, 1987 (IRRI, Manila), Drought Tolerance in Cereals in November, 1988 (Cairo, Egypt) and Rapid Propagation of Fast-growing Woody Species in November 1989 (Paris).

The idea to organize the present workshop was put forward at the 1987 CASAFA meeting in Warsaw, Poland. The suggested topic, *Control of Grassy Weeds in Tropical Subsistence Farming Systems*, brought together two of the Brundtland Commission Recommendations, namely sustainable agriculture and biological alternatives to pesticides.

Modern agriculture is impossible without efficient control of diseases, pests and weeds, and since the beginning of agriculture, some 10,000 years ago, farmers have been confronted with these same problems. For many centuries removal of weeds by traditional cultivation methods (such as ploughing, hoeing or by hand) has been both time- and labour-consuming. In the middle of 20th Century the introduction of growth regulators and herbicides that interfere with the growth of weeds or kill them was a revolutionary development. Chemicals and modern machinery made large-scale weed control possible at relatively low cost. Nowadays, herbicides are by far the most used chemicals in plant protection because they are both easy to apply and very effective. It is understandable why weed control is a great problem in developing countries: weeds grow more vigorously and regenerate more quickly in tropical areas than in temperate zones because of the higher temperatures and increased light intensity. However, much more is known about weed control techniques in the developed rather than in developing countries, but the methods of control used by farmers in the former are often unsuited to the physical, social and economic conditions of developing countries. Among the most destructive weeds are many grassy species, such as Bermuda grass, barnyard grass, Johnson grass, and cogon grass (alang-alang). They occur in every major agricultural area in the warmer regions of the world. As rotations become shorter in traditional tropical farming systems due to population pressure, the nutrients become depleted in the soil and as chemical control is beyond the means of the producers, grassy weeds become an impediment to sustainability.

With respect to the use of chemicals, there is another aspect which causes concern, and there is an increasing reluctance to depend too much on the use of chemicals for two reasons. First, the use of chemicals may have side-effects on other organisms and cause environmental hazards, and secondly there is the problem of development of resistance to herbicides in target plants. The burden of chemicals on non-target organisms and the environment should be made as low as possible, and the chemical selection pressure on the crops should be reduced in order to avoid herbicide resistance.

There are several ways to achieve this:

1. Optimal use of cultivation practices in strategies for weed control. The success of large-scale use of herbicides has decreased emphasis on the need to obtain detailed knowledge of weed biology. There are, nevertheless, many situations in which biological data on weed species can be of direct practical value to weed control.

2. The development of biological control. The possibility of utilizing organisms that attack weed species to achieve biological control is currently receiving much attention. According to Roberts (1982), insects have been employed against some 86 weed species, and measurable success obtained with 19 of them. Pathogenic fungi are increasingly being considered as control agents. For example, the introduction of *Puccinia chondrillina* (a rust) from Italy into Australia provided successful control of skeleton weed (*Chondrilla juncea*). The fungus *Colletotrichum gloeosporioides* has been applied successfully in Arkansas, USA, as a mycoherbicide against northern jointvetch (*Aeschynomene*) in rice and soyabean fields. Another biological agent is the grass carp, a freshwater fish from China, potentially valuable for water-weed control, particularly in situations where chemical herbicides cannot be used. So far, however, biological control of weeds is feasible only in a minority of cases.

3. Agronomic measures and adequate soil management can contribute to the reduction of noxious weed populations.

4. Finally a shift towards the use of more selective chemicals may reduce the effects on non-target organisms and the environment.

In this book, after two introductory reviews of the grassy weeds problem in the tropics, which also includes physiological and ecological aspects, attention is focused on the achievements and prospects of biological control, and on the state of affairs with respect to chemical control, including the safe use of pesticides on grassy weeds. A chapter then follows on the important subject of herbicide resistance – this chapter was commissioned specially for the book and was not presented as a paper at the workshop. These reviews are followed by presentations on integrated weed control in various important crops or groups of crops, namely in rice, cereal crops, forage crops and mixed crops. The volume concludes with regional overviews by participants from or with experience in tropical countries, and recommendations to ICSU, CASAFA and the agricultural community at large.

Acknowledgements

I should like to express thanks to my colleagues on the programme committee:

J.-P. Aeschlimann, CSIRO Biological Unit, Montpellier, France; H.C. Evans, International Institute of Biological Control (an Institute of CAB International), Ascot, UK; T.R. Roberts, Shell Research Limited, Sittingbourne, UK; and W. Sombroek, International Soil Research Institute, Wageningen, The Netherlands.

Special thanks are due to F.W.G. Baker who agreed to be secretary of the programme committee and helped organize the proceedings. Valuable advice was received from J. van der Heide, W. Joenje, L.'t Mannetje and P.J. van Rijn (Netherlands) and L. Knutson (Italy).

Finally, I wish to express our great appreciation to the Director of the

International Centre of Insect Physiology and Ecology, T.R. Odhiambo, and to the Deputy Director, P.B. Capstick, for hosting our workshop at ICIPE and to the staff, particularly Mrs R.P. Ortega, at ICIPE.

Reference

Roberts, H.A. (1982) The biology of weeds, In: *Weed Control Handbook* 1.

Chapter 1
Grassy Weeds – A General Overview

P.J. Terry

Tropical Weeds Unit, University of Bristol,
AFRC Institute of Arable Crops Research,
Long Ashton Research Station, Long Ashton,
Bristol, BS18 9AF, UK

Introduction

There is a clear botanical distinction between grasses (family Gramineae or Poaceae) and sedges (family Cyperaceae) but the similarity in the vegetative morphology of the two families can make them difficult for the untrained observer to distinguish. Hence the sedges are commonly mistaken for grasses because they generally have slender, elongated, parallel-veined leaves. For the purpose of this chapter, the sedges are considered to be grass-like or grassy weeds.

The Gramineae family is the fourth largest in the plant kingdom comprising approximately 9,000 species grouped into about 650 genera (Heywood, 1978). Grasses are widely dispersed (Willis, 1966), from polar regions to the equator and from mountains into the sea. They are ecologically the most dominant and economically by far the most important family (Heywood, 1978). It has been estimated that grasses are the principal component in 20% of the earth's cover of vegetation (Heywood, 1978) and that grasslands occupy a third of the world's land surface (Schantz, 1954). Grasses provide all the cereal crops, most of the world's sugar and grazing, and materials for construction and roofing. The grasses are a successful family for three reasons: '(a) their adaptability to changeable environments; (b) their ability to co-exist with grazing herbivores and man; and (c) their possession of a distinctive life form in which fidelity to a single architectural scheme is balanced by the endless ingenuity of its variations' (Clayton and Renvoize, 1986). Survival attributes of particular value in grasses are the vegetative shoot meristem situated near or just below the soil surface for a large part of the growing season, coupled with

Table 1.1. The most serious weeds in the world and distributions of the grasses and sedges (after Holm *et al.*, 1977).

Ranking	*Species**	*Type***	No. of countries	No. of crops
1	*Cyperus rotundus* L.	PS	92	52
2	*Cynodon dactylon* (L.) Pers.	PG	80	40
3	*Echinochloa crus-galli* (L.) P. Beauv.	AG	61	36
4	*Echinochloa colona* (L.) Link	AG	60	35
5	*Eleusine indica* (L.) Gaertn.	AG	60	46
6	*Sorghum halepense* (L.) Pers.	PG	53	30
7	*Imperata cylindrica* (L.) Raeuschel	PG	73	35
8	*Eichhornia crassipes* (Mart.) Solms	PAq		
9	*Portulaca oleraceae* L.	AD		
10	*Chenopodium album* L.	AD		
11	*Digitaria sanguinalis* (L.) Scop.	AG	56	33
12	*Convolvulus arvensis* L.	PD		
13	*Avena fatua* L. and relatives	AG	55	20
14	*Amaranthus hybridus* L.	AD		
15	*Amaranthus spinosus* L.	AD		
16	*Cyperus esculentus* L.	PS	30	21
17	*Paspalum conjugatum* Berg.	PG	30	25
18	*Rottboellia cochinchinensis* (Lour.) W.D. Clayton	AG	18	28

*Where possible, grass species names and authorities are as cited in Clayton and Renvoize (1972)

**A = annual, Aq = aquatic, D = dicot, G = grass, P = perennial, S = sedge

the ability to produce new tillers, thus conferring a resistance to grazing, slashing and fire. It is hardly surprising that such an extensive and versatile family contains many important crop and weed species.

The Cyperaceae family (sedges) consists of about 4,000 species in approximately 90 genera. Sedges are distributed in all parts of the world, especially in damp, wet or marshy regions of the temperate and subarctic zones (Heywood, 1978). Although some species are useful, sedges are of marginal economic importance as crops. However, several genera, including *Cyperus, Fimbristylis, Kyllinga, Mariscus* and *Scirpus*, contain some important species of weeds which are of considerable economic importance in terms of crop loss and costs of control.

Of approximately 250,000 plant species in the world, fewer than 250 are universally considered to be weeds (Holt, 1988). Approximately 71% of

Table 1.2. Grass and sedge weeds of secondary importance in the world (after Holm *et al.*, 1977).

Grasses	
Annual	*Perennial*
Cenchrus echinatus L.	*Axonopus compressus* (Sw.) P. Beauv.
Dactyloctenium aegyptium (L.) Willd.	*Brachiaria mutica* (Forssk.) Stapf
**Digitaria ciliaris* (Retz.) Koel.	*Digitaria abyssinica* (A. Rich.) Stapf
Ischaemum rugosum Salisb.	*Leersia hexandra* Sw.
Leptochloa panicea (Retz.) Ohwi	*Panicum maximum* Jacq.
**Leptochloa chinensis* (L.) Nees	*Panicum repens* L.
Lolium temulentum L.	*Paspalum dilatatum* Poir.
Setaria verticillata (L.) P. Beauv.	*Pennisetum clandestinum* Chiov.
Setaria viridis (L.) P. Beauv.	*Pennisetum pedicellatum* Trin.
	Pennisetum polystachion (L.) Schult.
	Pennisetum purpureum Schumach.
	Phragmites australia (Cav.) Steud.
	Phragmites karka (Retz.) Steud.
Sedges	
Annual	
Cyperus difformis L.	
Cyperus iria L.	
**Fimbristylis dichotoma* (L.) Vahl	
**Fimbristylis miliacea* (L.) Vahl	

*annual or perennial habit

the world's weeds occur in eight plant families, and over 50% of these weed species are found in only two families, Compositae and Gramineae (Holt, 1988).

In this chapter, grasses and sedges are considered in terms of their weedy characteristics and their relationships with crops.

Important species of grassy weeds

The grassy weeds are highly conspicuous, and often significant, components of the flora of virtually every cropping system in the world. They can be found in highly developed and primitive agro-ecosytems, and in all temperate and tropical crops. In a worldwide survey of weeds, Holm *et al.*

(1977) ranked and described the most important species. Of the 18 most serious weeds, ten are grasses and two are sedges (Table 1.1).

Holm *et al.* (1977) went on to describe a second group of 72 weeds which are widespread and troublesome but which are perhaps rather less important than those listed in Table 1.1. They include 22 grasses and four sedges (Table 1.2). Whether or not one agrees with the ranking and relative importance of the species assigned by Holm *et al.* (1977), it is clear from their survey that Gramineae is the most important family of weeds. Illustrations of 20 major grassy weeds important in the tropics are shown in Figure 1.1 at the end of this chapter.

Characteristics of success for grassy weeds

Weeds need a combination of attributes if they are to be successful. Grasses and sedges are no different from other weeds in this respect and some of the characteristics for their success are presented below. One of the keys to the control of weeds is to prevent the expression of these attributes.

Dispersal

Weeds can be dispersed in several ways, as is shown by the following examples:

Wind the feathery seeds of *Imperata cylindrica* can be blown over large distances though the average flight at inflorescence level is reputed to be 15 m (Holm *et al.*, 1977).
Water many annual and perennial grasses and sedges grow in rice paddies and on the banks of irrigation channels.
Animals the burrs of *Cenchrus echinatus* and the barbed bristles of the inflorescence of *Setaria verticillata* enable these grasses to become attached to fur, feathers and skins of animals and the clothes of man. Seeds of *Cynodon dactylon* survive ingestion by cattle and can be dispersed in this way.
Man the seeds and perennating organs of many weeds are dispersed on farm machinery, such as muddy tractor wheels and cultivation implements. The dispersal of wild oats (*Avena* spp.) through the cereal-growing areas of Kenya can be largely attributed to unclean combine harvesters. Weeds with seeds of similar size to the crop can be dispersed in crop seed, e.g. wild oats in wheat, *Rottboellia cochinchinensis* and *Ischaemum rugosum* in rice. Weedy mimics can be transplanted by mistake with the crop, e.g. wild rice. (*Oryza barthii* A. Chev., *O. rufipogon* Griff., etc.) in rice. Many weeds owe their existence in some countries to transportation in crop seed

across international boundaries. *O. rufipogon*, for example, became established in Swaziland after being introduced from the southern USA.

Dormancy

Seeds and perennating organs which can remain viable for long periods and have pronounced dormancy enable weeds to survive in the soil when conditions are not suitable for growth. Dormancy mechanisms in the rhizomes and stolons of many perennial grasses and sedges have unquestionably contributed to their success as weeds. Lateral buds on rhizomes are suppressed by growth of the rhizome apex until death of the apex or fragmentation of the rhizome. In one of the most successful weeds, *Cyperus rotundus*, apical growth of the rhizome suppresses growth of tubers on that rhizome and apical buds on the tubers suppress the growth of lateral buds. Hence, a hectare of land containing 10 million *Cyperus* tubers can have ten times as many viable, dormant buds, giving the weed phenomenal potential for regeneration.

Grass seeds are not noted for their longevity. Even the most persistent of grass seeds are not viable beyond 6–9 years, with a death rate of 70% of the population per year. Seeds of many serious grassy weeds survive for only short periods, e.g. about one year for *Bromus* spp. and three years for *Rottboellia cochinchinesis*, giving scope for control if seed production can be prevented for those lengths of time. This is in marked contrast to many dicot weed seeds which can survive for as long as 50 years in poppies (*Papaver* spp.) or 20 years in *Striga* spp.

There are many ways in which weeds exploit dormancy to ensure their survival. Successful management of weeds often requires a knowledge of this attribute, even if there is an incomplete understanding of dormancy mechanisms.

Germination and establishment

Weeds are usually most vulnerable to control immediately after the germination of their seeds. Damage to or destruction of the young shoot at this stage often leads to death of the seedling. For perennial plants with relatively large reserves of food in their perennating organs, death of the young shoot is less important when there are dormant buds held in reserve. Successful weeds must therefore have the capacity to recover from damage during this critical time and/or hasten establishment by means of rapid growth. In this respect, grassy weeds are no different from other weed species.

Weeds that are efficient in their capture of resources (water, nutrients, light, etc.) have higher relative growth rates which is a helpful attribute for plants growing in disturbed soils. Photosynthetic efficiency is also a useful attribute for a weed to possess. Thus weeds with C4 metabolism are likely to be more competitive than crops or weeds with the less efficient C3 metabolism. It is probably no coincidence that 42% of the world's worst weeds cited by Holm *et al.* (1977) have C4 photosynthesis and 78% of the top 18 weeds are C4 (Holt, 1988). However, most of the world's major crops have C3 photosynthesis (Holt, 1988), notable exceptions being maize, sorghum and sugarcane. In practice, the advantage of C4 photosynthesis depends on environmental conditions; water- and heat-stressed environments favour the C4 species. Regardless of photosynthetic mechanisms, the leaf architecture of grass-like plants enables them to benefit from increased light penetration into their canopies and they have a high biomass in a relatively small space (Clayton and Renvoize, 1986), which could be advantageous to grasses growing in less efficient broadleaved crops.

Adaptability and plasticity

Weeds are typically found in disturbed sites where there are wide variations in temperature, moisture, space and exposure. Weeds which can tolerate these conditions are best able to survive in agro-ecosystems and their adaptibility and plasticity are major factors contributing to their success. Harper (1977) refers to plasticity as the capacity for phenotypic adjustment in morphological characteristics and physiological responses under changing conditions. It is a character well recognized in all the grass and sedge weeds listed in Table 1.1. A single plant of *Avena fatua*, for example, growing in ideal conditions can produce 2,000 seeds but in extreme, adverse conditions an emaciated plant may still produce, say, five seeds which are sufficient to produce another generation. At a single site, *Imperata cylindrica* can tolerate drought, waterlogging, fire, cultivation and short-term shade, unlike most other plants. Adaptability is well demonstrated by *Cyperus rotundus* which occurs in over 50 different crops and grows 'with reckless abandon in almost every soil type, elevation, humidity, soil moisture and pH, and can survive the highest temperatures known in agriculture' (Holm *et al.*, 1977).

Propagation

Successful weeds need to be good at propagating themselves. High seed production is an advantage but is not a character to separate grassy weeds from other species. Single plants of *Echinochloa crus-galli* are reported to produce 2,000 – 40,000 seeds and in a weedy field could produce more than one tonne of seed per hectare (Holm *et al.*, 1977).

Effective perennation, one of the attributes of an 'ideal weed' (Baker,

1974), has contributed to the success of four grasses and two sedges in the world's 18 worst weeds (Table 1.1) and many other species. A single tuber of *Cyperus rotundus* has produced 99 tubers in 90 days in India (Rao, 1968) and tuber populations of 10 million/ha are possible. In one hectare of land, Soerjani (1970) estimated that *Imperata cylindrica* can produce six tonnes of rhizomes and 4.5 million shoots with a mass of ten tonnes.

The perennating habit of many grassy weeds has ensured their success in smallholder cropping systems. Cultivations by hand-tools and animal-drawn implements are often not adequate to bring rhizomes to the soil surface where they can be removed or exposed to desiccation. What is more likely to happen is that rhizomes are broken into smaller viable fragments and the weed problem is exacerbated. Where infestations are severe, smallholders will often abandon land rather than waste their time with cultivations.

Problems caused by grass and sedge weeds

Grass and sedge weeds can affect crops in many adverse ways, most of which are no different from weeds in general.

Competition

Weeds have a direct effect on crop yields when they compete with the crop for nutrients, moisture and light. Zimdahl (1980) cites 586 literature sources in his review of weed–crop competition but many other references can be found. Grasses and sedges usually compete with crops as part of a mixed weed flora. There are, however, reports of research on the competitive effects of individual grass weeds.

Smith (1968), studying the interaction of *Echinochloa crus-galli* and rice, found that weed populations of 11, 54 and 269 plants/m^2 caused yield reductions in optimum rice stands (334 plants/m^2) of 25, 59 and 79%, respectively. At lower crop densities, the yield reductions were higher, confirming that this grass weed was more competitive in sparse rice stands than in vigorous ones.

Knake and Slife (1962) showed that yields of maize were reduced by 7, 12 and 24% with *Setaria faberii* Herrm. at densities of 3, 20 and 177 plants/m of crop row, respectively. In sorghum, competition from *Pennisetum glaucum* (L.) R.Br. (*Setaria lutescens* (Weigel) F.T. Hubbard) for six weeks reduced grain yield by 28% and, in a full season's competition, by 53% (Hurst, 1966).

The level of competition between weeds and crops varies considerably with factors such as soil fertility, soil moisture and crop variety. For example, it has been shown that *Lolium multiflorum* causes a greater yield reduction in semi-dwarf varieties of wheat than in tall varieties (Appleby *et al.*, 1976). The effect of soil fertility is variable, due no doubt to the fact that both crop and weeds can benefit from it. Zimdahl (1980), citing other authors, tabulates results showing that annual grasses in rice at high levels of fertility reduce crop yields of cultivars H-4 and Shiranuhi but for IR-8 the yield reduction is greater at normal soil fertility. Periodicity of weed germination is another factor affecting weed–crop competition: late germinating weeds (as grass weeds often are) are likely to have a greater effect on late sown crops and long duration varieties than weeds that germinate early in the season.

Allelopathy

Allelopathy, the chemical interaction between one plant and another, has been a subject of some controversy but it is now generally accepted that certain weeds have the ability to adversely affect the growth of some crops. Grass weeds with allelopathic properties include *Cynodon dactylon, Sorghum halepense* (Horowitz and Friedman, 1971) and *Imperata cylindrica* (Eussen and Wirjahardja, 1973), but many more are suspected. Residues of killed weeds can also damage crops by allelopathy and this effect may extend beyond the season when weed control took place.

Interference with harvesting

Spiny burrs on the seeds of *Cenchrus echinatus* and *C. biflorus* Roxb. are a painful deterrent to the hand-harvesting of many tropical crops. Some weeds, such as *Setaria verticillata* in wheat, produce such an entwined mat of bristly inflorescences over the crop canopy that mechanical harvesting has to be abandoned.

Contamination of crops

The value of crops grown for seed is considerably reduced when contaminated by weed seeds. Legislation in many countries is designed to ensure that weeds such as *Avena fatua* in cereals are not spread through the use of impure crop seed. Contamination of rice grain by the grain of red rice (*Oryza rufipogon*) necessitates longer processing to polish away the red pigment, thereby reducing the value of the crop.

Poisoning of man and livestock

Lolium temulentum has been known since biblical days as a cereal weed that produces seeds that poison man and livestock. Poisoning occurs when contaminated cereal grain is eaten. The toxin appears to be produced from a fungus, *Endoconidium temulentum* Prill & Dell., which is present in the grain of the weed.

Loss of irrigation water

An indirect effect of weeds on crops occurs when water flow in irrigation channels is reduced by weeds, such as grasses, reeds and sedges, growing in the channels or extending from the banks. Weeds can also increase water loss from reservoirs, lakes and canals by means of evapotranspiration. In India, for example, it has been observed that one hectare of *Typha* spp. (a grass-like genus in the Typhaceae) transpires 80 million litres of water per year, enough to irrigate 4.6 ha of rice or 11 ha of wheat (Gupta, 1976).

Alternative host for crop pests

It is hardly surprising that graminaceous weeds can be alternative hosts for pests of graminaceous crops or that they and other plants growing in association with the crop can provide sites and reservoirs of infection. Many examples can be found in the literature, including those of Bendixen (1983) who listed the numbers of arthropod and nematode pests of crops which occur on some of the world's worst weeds listed by Holm *et al.* (1977). *Cynodon dactylon*, for example, is reported to be an alternative host for 11 arthropod and 12 nematode pests. Examples of grassy weeds which are alternative hosts of rice pests are shown in Table 1.3.

Table 1.3. Grass weeds which are alternative hosts of rice pests.

Weed	*Pest*	*Type of pest*
Leersia hexandra	rice leaf roller	insect
	brown plant hopper	insect
	rice green looper	insect
Arundinella leptochloa (Nees) Hook.f.	yellow rice borer	insect
Echinochloa colona	rice blast	fungus
Axonopus compressus	rice sheath blight	fungus
Eleusine indica	ragged stunt virus	virus
Oryza longistaminata A. Chev. & Roehr.	yellow mottle virus	virus
Many weeds	rice root nematode	nematode

Grass weeds can also be hosts of *Striga* spp., an important genus of parasitic weeds which cause considerable damage to cereal crops in many tropical countries. In the USA, *Digitaria sanguinalis* (L.) Scop. is the host for *Striga asiatica* (L.) O. Ktze., a parasite of maize, but other grasses are known or suspected to be alternative hosts of *S. asiatica, S. hermonthica* (Del.) Benth. and *S. aspera* (Willd.) Benth. in Africa.

Weeds can also provide sustenance and cover for pests such as rats and grain-eating birds that plague many farms in developing countries.

Associations of grassy weeds with crops

Grass and sedge weeds are so diverse in their biology that there is no simple explanation as to why they contribute so much more to the weed flora than other families. It is probably no coincidence that a family that can produce so many crops can also provide the weeds to infest them.

Genetic relationships

Nearly all field and garden crops and some tree crops have weed races that belong to the same biological species as the crop. For example, there are weed races of barley, maize, rice, sorghum and wheat which can, if genetically compatible, hybridize with the crop. Harlan (1982) reviewed some aspects of the relationship between crops and weeds and used the genetic interactions between grain sorghum and its weed races as a model of this phenomenon. Through hybridization, weedy sorghums are produced which closely resemble the crop but shatter before harvest, a characteristic which has distinct survival advantages for a weed but which is undesirable for a crop. It has been observed that such weeds can mimic the sorghum in which they grow; for example, if the crop has loose, open panicles, so too will the weed. The mimicry is largely genetically controlled but is assisted by the fact that the mimics are less likely to be weeded out of the crop. Mimicry by weeds is found in several grass crops, including maize, millets (barnyard, foxtail and pearl), oats, rice and rye.

Harlan (1982) also reviewed cultural relationships between crops which show how grass weeds become important in cereal crops. He noted that the selection pressure of hand-weeding can produce mimics of unrelated species. For example, *Echinochloa crus-galli* var. *oryzicola* (Vasig.) Ohwi closely resembles the crop in its vegetative phase. By the time that the distinctive inflorescence appears, the weed has damaged the crop. Such mimicry by unrelated weeds also occurs in other crops such as rye and oats.

Table 1.4. Examples of grass and sedge weeds associated with crop production.

Crop production system	*Associated species*
Flood irrigated crops, especially rice	*Echinochloa* spp., *Ischaemum rugosum*, *Leptochloa* spp., wild rice (*Oryza* spp.), *Cyperus difformis*, *C. iria*, *Fimbristylis* spp.
Mangrove rice	*Paspalum vaginatum* Sw.
Slash-and-burn agriculture	*Imperata cylindrica*
Minimum and zero tillage	*Cyperus rotundus* and other perennial sedges, *Cynodon dactylon* and other perennial grasses
Mowing and repeated cutting, e.g. in citrus plantations and along banks of irrigation channels	*Cynodon dactylon, C. nlemfuensis* Vanderyst and other prostrate species
Shady plantation crops	*Paspalum conjugatum*, *Ottochloa nodosa* (Kunth) Dandy, *Setaria homonyma* (Steud.) Chiov.
Small-grain cereal crops	wild oats (*Avena* spp.), *Bromus* spp., *Phalaris* spp., *Setaria verticillata* and many other annual grasses
Irrigated, intensively cultivated nurseries and horticultural crops	*Cyperus rotundus*
Grass crops managed by the use of selective herbicides (e.g. 2, 4–D and MCPA in wheat, atrazine in maize and sugarcane, etc.)	many grass weeds which are tolerant of the same herbicides as the crops (e.g. *Rottboellia cochinchinensis* in maize and sugarcane)

Crop production relationships

The production of a crop requires a particular set of conditions in order to maximize growth and yield. These conditions include time of cultivation, time of planting, spacing, irrigation, fertilizer application and use of herbicides. Repeated growing of the same crop and use of the same or similar production practices imposes a selection pressure on any weeds that are present. The result is that weeds tolerant of these practices emerge as dominant species. These can be difficult to control without changing the crop and cultural conditions. Grass and sedge weeds have certainly exploited the niche created by the conditions needed to grow a wide range of crops (Table 1.4). Whilst it is easy to advise resource-poor farmers how to avoid weed problems by changing their methods of crop production, it is often very difficult for them to accept the advice. Many social and economic factors such as lack of cash or credit to purchase inputs, shortage of labour or equipment, dietary requirements, conservative attitudes towards change and aversion to taking any risks dictate what these farmers can do.

Weed management

Weed control is often used as a euphemism for weed management but the two are not necessarily the same. Management implies that some long- or short-term strategy is employed to reduce weed infestations. It involves a degree of foresight and predictability in order to plan treatments and integrate them into cropping systems. This is not necessarily how farmers in developing countries tackle a weed problem. They are more likely to wait until they see the weeds and then 'control' them by cultivations, hand-pulling, etc. Whilst this is an immediate remedy for a perceived problem, it may not provide the long-term solution. Difficult weeds, such as the perennial grasses and sedges, need to be managed over an extended period if they are to be satisfactorily eradicated.

The management of weeds is very much related to the crops they infest and the techniques used to grow the crops. The types of weed control can be divided into physical, cultural, chemical and biological methods. No single method will solve all weed problems but may exacerbate the problems caused by some species. Indeed, the proliferation of grass and sedge weeds from within a mixed weed flora is a common phenomenon for some of the reasons given above. Hence the need for a range of options for managing weeds which can be integrated into the various ways of producing crops. The management will differ between crops, sites, weeds present and the level of farming. Methods appropriate to the low-input smallholder will differ from those of the capital intensive estates. However, the principles of weed management remain the same and are based on three fundamental (and broad) parameters; the cropping system, the biology of the weed and the resistance of the weed to control. A weed can become a problem with any permutation of these parameters. A species which is easy to control can still present a problem if it is a profuse seed producer and/or germinates over an extended period. By contrast, *Bromus* spp. in the UK have meagre dormancy and poor seed longevity (biological characteristics not usually associated with a successful weed) but they are difficult to control and are well adapted to continuous cereal production, thus becoming serious weeds.

A problem that has yet to manifest itself in any serious way in the tropics is the development of herbicide resistance in weeds. It is only a matter of time before the predisposing factors, such as continued use of one herbicide group, provide the necessary selection pressures to produce resistance in hitherto susceptible species.

Grassy weeds in the future

There is little cause for optimism that farmers in developing countries will cope any better with grassy weeds than those in the developed world where, despite the use of relatively advanced technology and high inputs, many grasses and sedges are still major weeds. It is certainly possible to learn from some of the mistakes made in Europe, America and elsewhere. For example, the developing countries could prevent the further importation and movement of weed seeds and propagules by rigorously enacting and enforcing legislation on quarantine and seed purity. It should also be possible to adapt some herbicide technology and integrate it into local conditions; for example, the use of supplementary cultivations to prevent the establishment of tolerant and resistant weeds. It is inevitable that methods of crop production and weed management in developing countries will change. It is equally certain that the weed flora will change in response to new selection pressures; easily controlled weeds will be replaced by more obstinate species. Grasses and sedges will become more important as weeds because 'improved' weed management tends to select for them. Herbicide resistance will become more evident, especially where weeds can complete two or more generations in a season. The general principles for minimizing the effects of these trends are known: e.g. rotation of crops, tillage systems and herbicides; use of herbicide mixtures; integration of chemical, mechanical and cultural methods of control. Unfortunately, we lack detailed knowledge of how best to apply the principles to specific problems.

The implementation of any improved weed management in developing countries requires a cadre of scientists and extension officers with the necessary training and expertise. Unfortunately, most countries have not addressed this need, with the result that they are unable to benefit from advances in weed science. This situation must be remedied if solutions are to be found to one of the most serious constraints to crop production in the developing world.

Acknowledgements

I am grateful to my colleagues G.W. Cussans, C. Parker and C.R. Riches for their constructive advice during the preparation of this paper.

References

Appleby, A.P., Olson, P.D. and Colbert, D.R. (1976) Winter wheat yield reduction from interference by Italian ryegrass. *Agronomy Journal* 68, 463–6.

Baker, H.G. (1974) The evolution of weeds. *Annual Review of Ecology and Systematics* 5, 1–24.

Bendixen, L.E. (1983) Another criterion in defining the world's worst weeds. *Proceedings of the 9th Asian–Pacific Weed Science Society Conference,* 176–80.

Clayton, W.D. and Renvoize, S.A. (1972) Gramineae (Part 3). In: Polhill, R.M. (ed.) *Flora of Tropical East Africa.* A.A. Balkema, Rotterdam, 451–898.

Clayton, W.D. and Renvoize, S.A. (1986) *Genera Graminium. Grasses of the World.* London, 389 pp.

Eussen, J.H.H. and Wirjahardja, S. (1973) Studies on alang-alang (*Imperata cylindrica* (L.) Beauv.) vegetation. *BIOTROP Bulletin* 6, Regional Centre for Tropical Biology, P.O. Box 17, Bogor, Indonesia.

Gupta, O.P. (1976) Aquatic weeds and their control in India. *FAO Plant Protection Bulletin* 24, 76–82.

Harlan, J.R. (1982) Relationships between weeds and crops. In: Holzner, W. and Numata, N. (eds) *Biology and Ecology of Weeds.* Dr. W. Junk Publishers, The Hague, pp91–6.

Harper, J.L. (1977) *Population Biology of Plants.* Academic Press, New York.

Heywood, V.H. (1978) *Flowering Plants of the World.* Oxford University Press, 335pp.

Holm, L.G., Plucknett, D.L., Pancho, J.V. and Herberger, J.P. (1977) *The World's Worst Weeds. Distribution and Biology.* The University Press of Hawaii, Honolulu, 609pp.

Holt, J.S. (1988) Ecological and physiological characteristics of weeds. In: Altieri, M.A. and Liebman, M. (eds) *Weed Management in Agro-ecosystems: Ecological Approaches.* CRC Press, Boca Raton, Florida, pp7–23.

Horowitz, M. and Friedman, T. (1971) Biological activity of subterranean residues of *Cynodon dactylon, Sorghum halepense* and *Cyperus rotundus. Weed Research* 11, 88–93.

Hurst, H.R. (1966) Competition of four weed species with grain sorghum. *Dissertation Abstracts* 1349–B, Kansas State University, USA.

Knake, E.L. and Slife, F.W. (1962) Competition of *Setaria faberii* with corn and soybeans. *Weeds* 10, 26–9.

Rao, J. (1968) Studies on the development of tubers in nutgrass and their starch content at different soil depths. *Madras Agricultural Journal* 55, 19–23.

Schantz, H.L. (1954) The place of grasslands in the earth's cover of vegetation. *Ecology* 35, 143–5.

Smith, R.J., Jr. (1968) Weed competition in rice. *Weed Science* 16, 252–4.

Soerjani, M. (1970) Alang-alang, *Imperata cylindrica* (L.) Beauv., pattern of growth as related to its problem of control. *BIOTROP Bulletin* 1, Regional Centre for Tropical Biology, P.O. Box 17, Bogor, Indonesia.

Willis, J.C. (1966) *A Dictionary of the Flowering Plants and Ferns* (7th Edn). Cambridge University Press, pp1214 + liii.

Zimdahl, R.L. (1980) *Weed–Crop Competition: A Review.* International Plant Protection Center, Oregon State University, Corvallis, Oregon, USA, 196pp.

Grassy weeds of importance in the Tropics

Fig. 1.1a *Avena fatua* L. 1. habit; 2. ligule; 3. spikelet attached to rachis, note long awns; 4. detached spikelet showing 'suckermouth'; 5. glume; 6. glume; 7. grain; 8. grain, cross section (reproduced, with permission of the University Press of Hawaii, from Holm *et al.*, 1977).

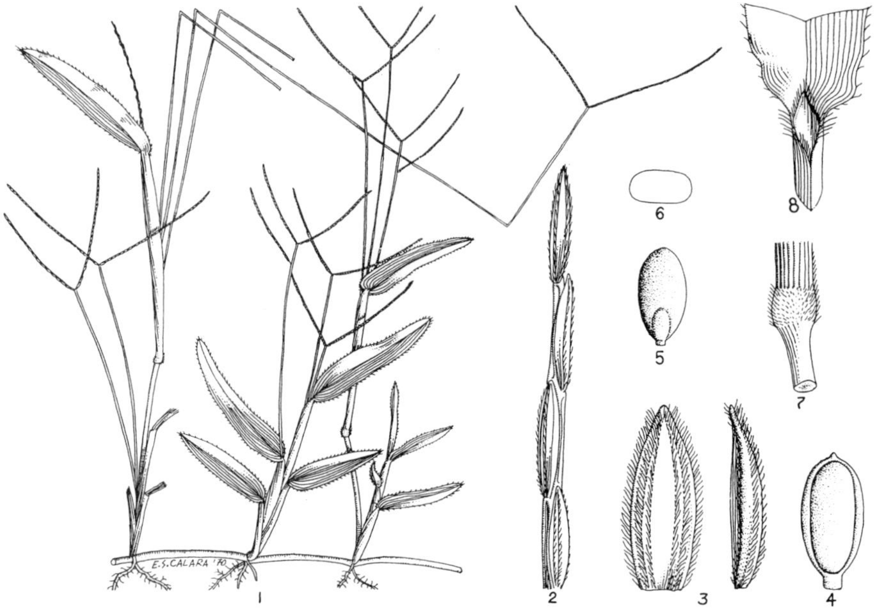

Grassy weeds of importance in the Tropics
Fig. 1.1b *Axonopus compressus* (Sw.) P. Beauv. 1. habit; 2. portion of spike; 3. spikelet, two views; 4. floret; 5. grain; 6. grain, cross section; 7. portion of culm with node; 8. ligule (reproduced, with permission of the University Press of Hawaii, from Holm *et al.*, 1977).

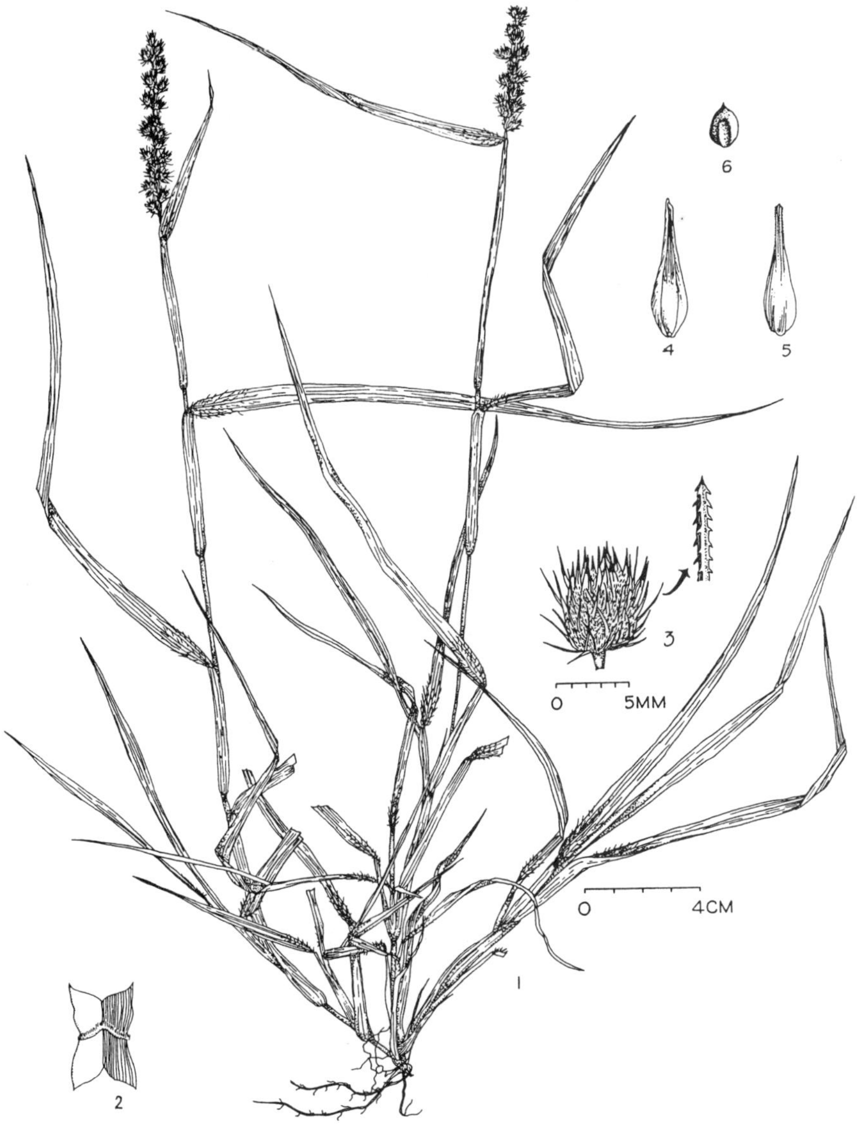

Grassy weeds of importance in the Tropics
Fig. 1.1c *Cenchrus echinatus* L. 1. habit; 2. ligule; 3. bur, with detail of ligule; 4, 5 spikelet, two views; 6. seed (reproduced, with permission of the University Press of Hawaii, from Holm *et al.*, 1977).

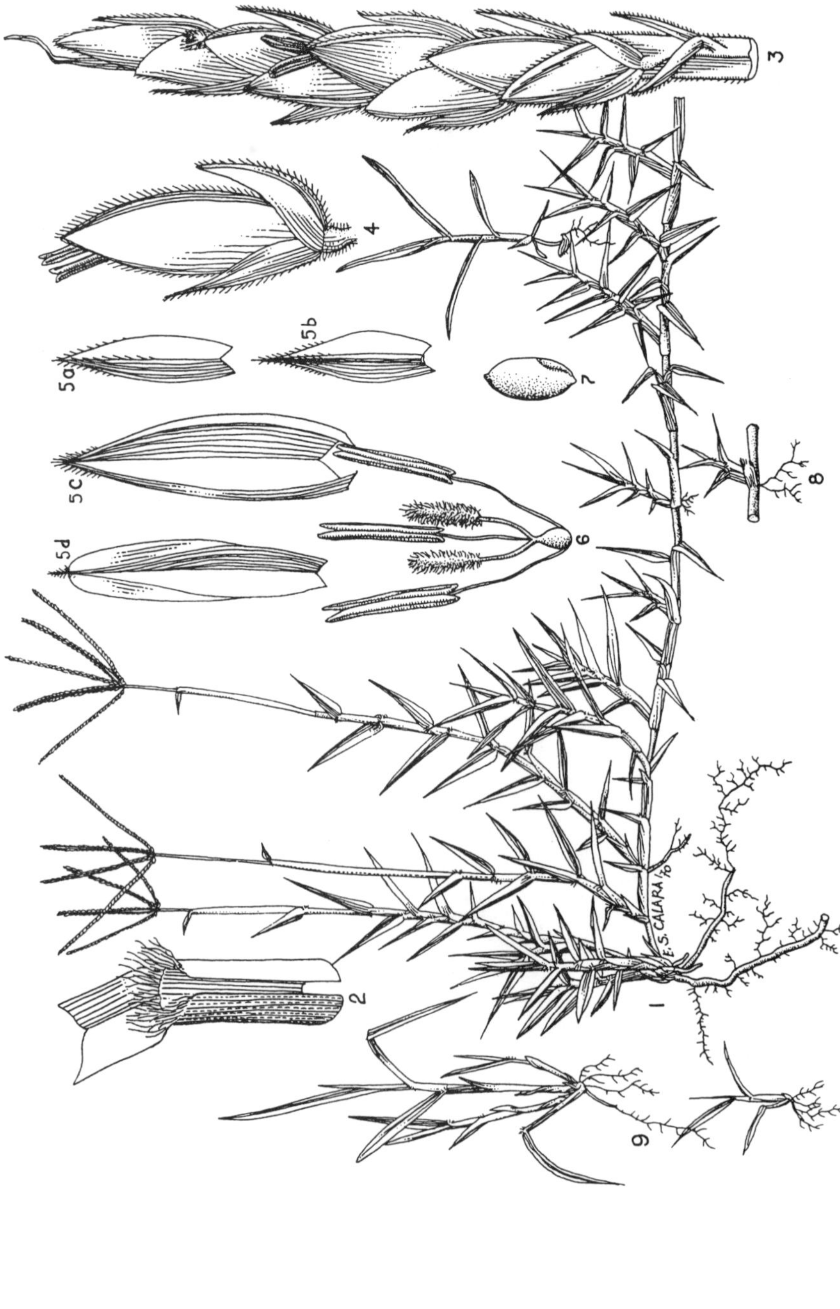

Grassy weeds of importance in the Tropics

Fig. 1.1d *Cynodon dactylon* L. Pers. 1. habit; 2. ligule; 3. portion of inflorescence; 4. flower; 5. *a–d*, bracts; 6. flower; 7. grain; 8. portion of culm; 9. seedling (reproduced, with permission of the University Press of Hawaii, from Holm *et al.*, 1977).

Grassy weeds of importance in the Tropics

Fig. 1.1e *Cyperus difformis* L. 1. habit; 2. portion of culm; 3. spikelet; 4. bract; 5. flower; 6. achene; 7. achene, cross section; 8. seedling (reproduced, with permission of the University Press of Hawaii, from Holm *et al.*, 1977).

Grassy weeds of importance in the Tropics

Fig. 1.1f *Cyperus esculentus* L. 1. habit; 2. spikelet; 3. bract; 4. seed, with closeup detail of surface reticulations; 5. seed, cross section (reproduced, with permission of the University Press of Hawaii, from Holm *et al.*, 1977).

Grassy weeds of importance in the Tropics
Fig. 1.1g *Cyperus iria* L. 1. habit; 2. spikelet; 3. achene; 4. achene, cross section; 5. detailed surface of reticulations; 6. seedlings (reproduced, with permission of the University Press of Hawaii, from Holm *et al.*, 1977).

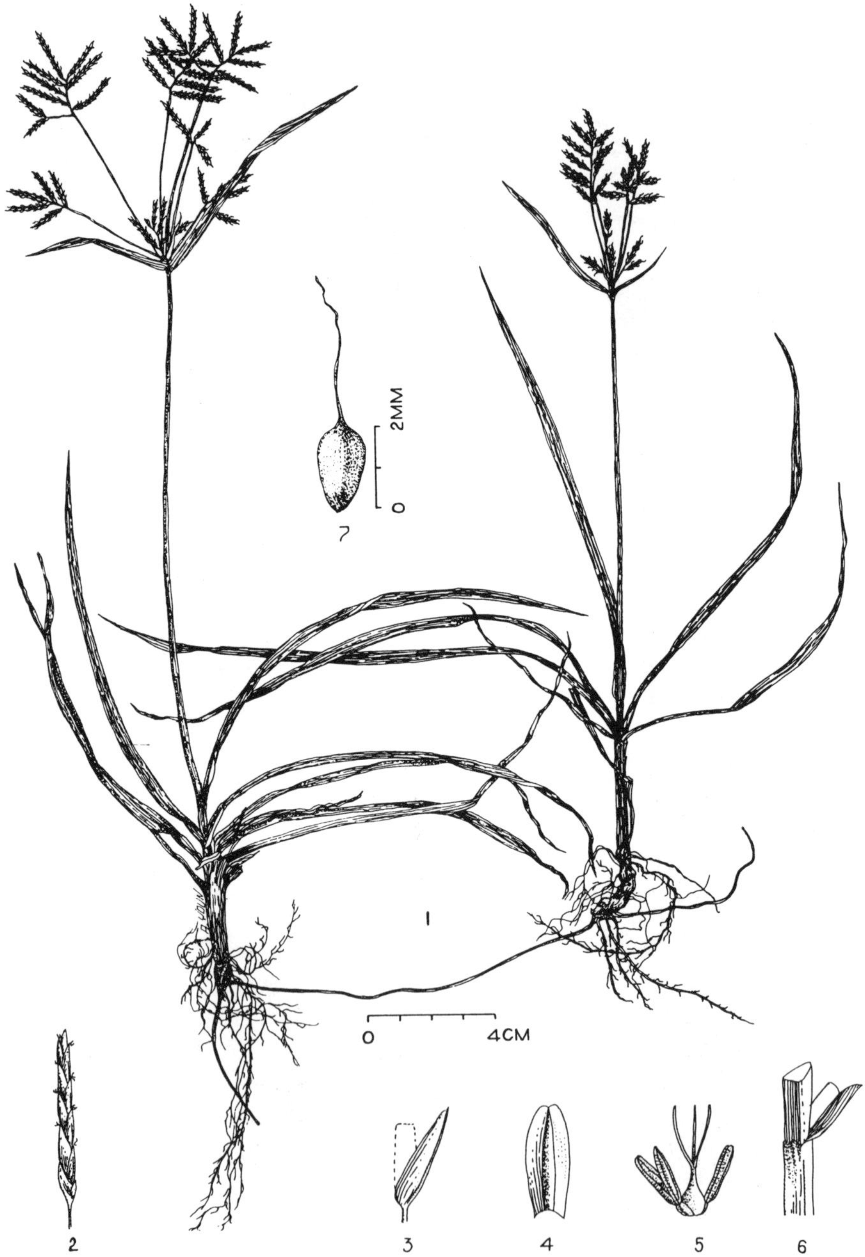

Grassy weeds of importance in the Tropics

Fig. 1.1h *Cyperus rotundus* L. 1. habit; 2. portion of inflorescence; 3. flower with glume; 4. glume; 5. flower, glume removed; 6. portion of leaf sheath and blade; 7. achene (reproduced, with permission of the University Press of Hawaii, from Holm *et al.*, 1977).

Grassy weeds of importance in the Tropics

Fig. 1.1i *Dactyloctenium aegyptium* L. Willd. 1. habit; 2. ligule; 3. spikelet; 4. glumes, upper and lower; 5. grain (reproduced, with permission of the University Press of Hawaii, from Holm *et al.*, 1977).

Grassy weeds of importance in the Tropics

Fig. 1.1j *Digitaria sanguinalis* L. Scop. 1. habit; 2. ligules; 3. portion of spike; 4. spikelet, front view; 5. spikelet, back view; 6. spikelet, bracts removed; 7. grain; 8. grain, cross section; 9. seedlings (reproduced, with permission of the University Press of Hawaii, from Holm *et al.,* 1977).

Grassy weeds of importance in the Tropics

Fig. 1.1k *Echinochloa crus-galli* L. P. Beauv. 1. habit; 2. ligule; 3. spikelet, back view; 4. spikelet, side view; 5. spikelet, back view; 6. flower, glumes excised; 7. flower; 8. grain; 9. grain, cross section; 10. seedling (reproduced, with permission of the University Press of Hawaii, from Holm *et al.*, 1977).

Grassy weeds of importance in the Tropics

Fig. 1.1l *Echinochloa colona* L. Link 1. habit; 2. inflorescence; 3. spikelet, ventral view; 4. spikelet, dorsal view; 5. floret; 6. bracts; 7. portion of culm; 8. portion of leaf base and blade; 9. grain; 10. grain, transverse section; 11. seedlings (reproduced, with permission of the University Press of Hawaii, from Holm *et al.*, 1977).

Grassy weeds of importance in the Tropics

Fig. 1.1m *Eleusine indica* L. Gaertn. 1. habit; 2. ligule; 3. spikelet; 4. bract; 5. flower; 6. seed, two views; 7. seed, cross section; 8. seedling (reproduced, with permission of the University Press of Hawaii, from Holm *et al.,* 1977).

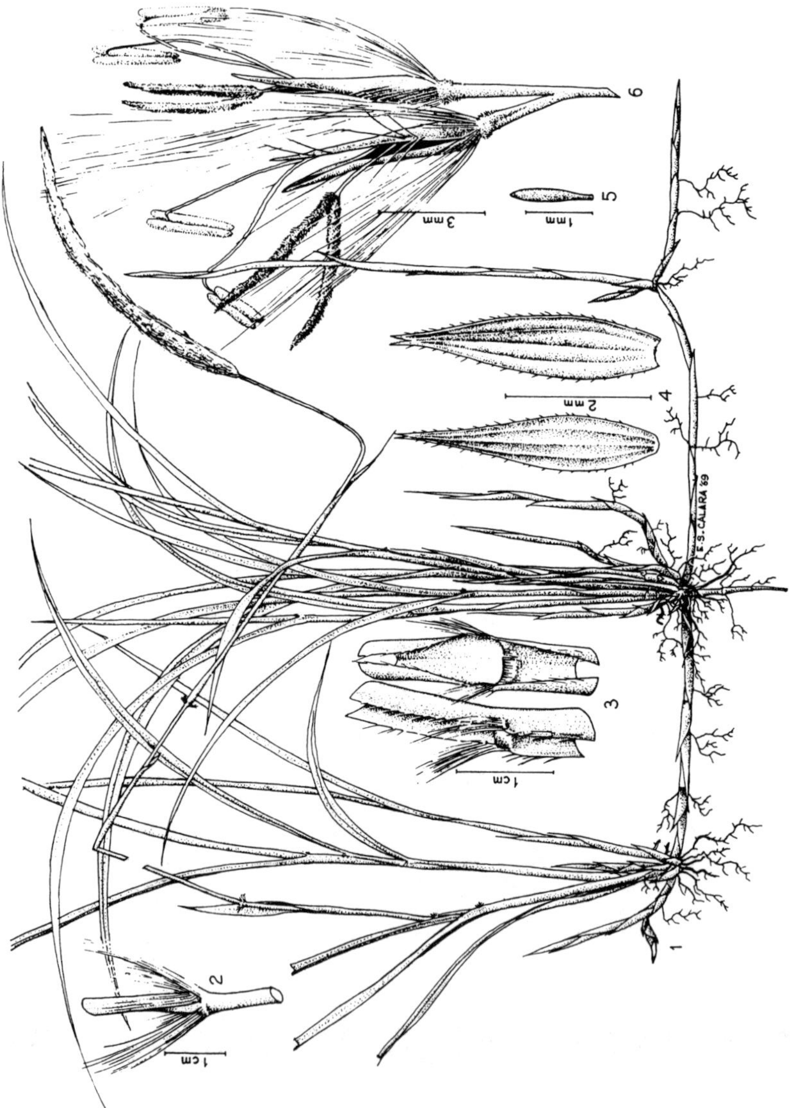

Grassy weeds of importance in the Tropics

Fig. 1.1n *Imperata cylindrica* L. Raeuschel 1. plant showing rhizome; 2. node with silky hairs; 3. ligule and portion of sheath and blade, ciliate; 4. lemma and palea; 5. grain; 6. spikelets (reproduced, with permission of the University Press of Hawaii, from Holm *et al.*, 1977).

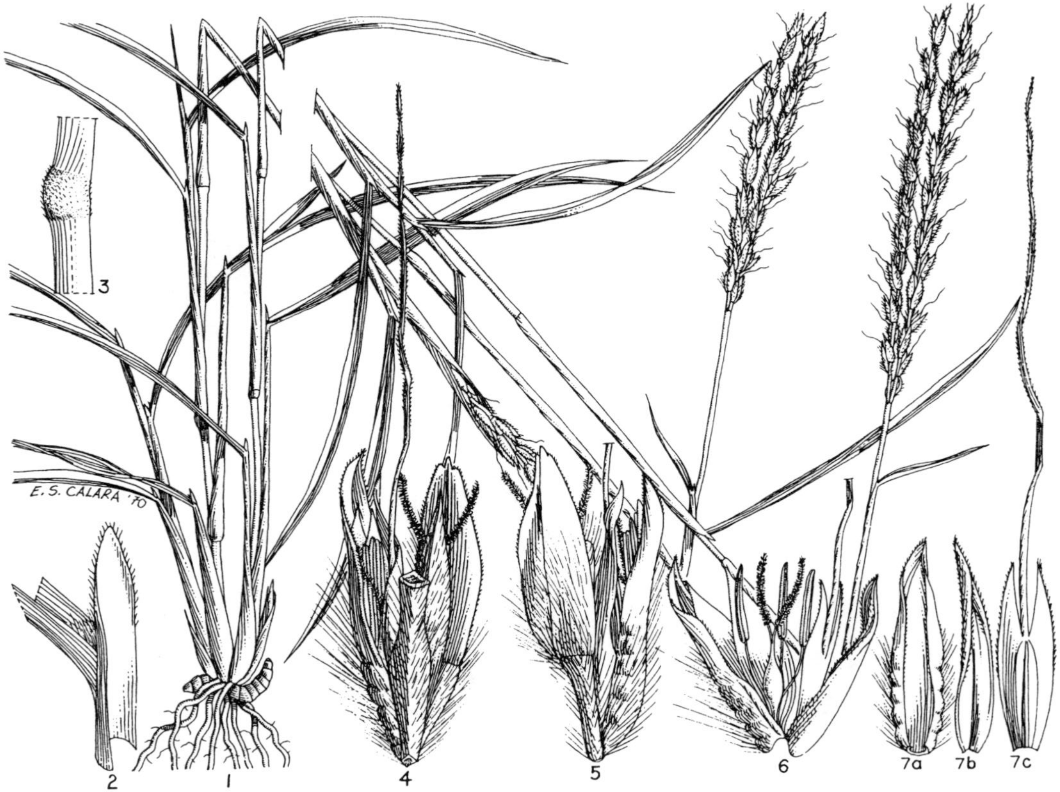

Grassy weeds of importance in the Tropics

Fig. 1.10 *Ischaemum rugosum* Salisb. 1. habit; 2. ligule; 3. portion of culm to show the tufted hairy node; 4. spikelet, front view; 5. spikelet, back view; 6. spikelet, partly opened; 7*a–c* bracts (reproduced, with permission of the University Press of Hawaii, from Holm *et al.*, 1977).

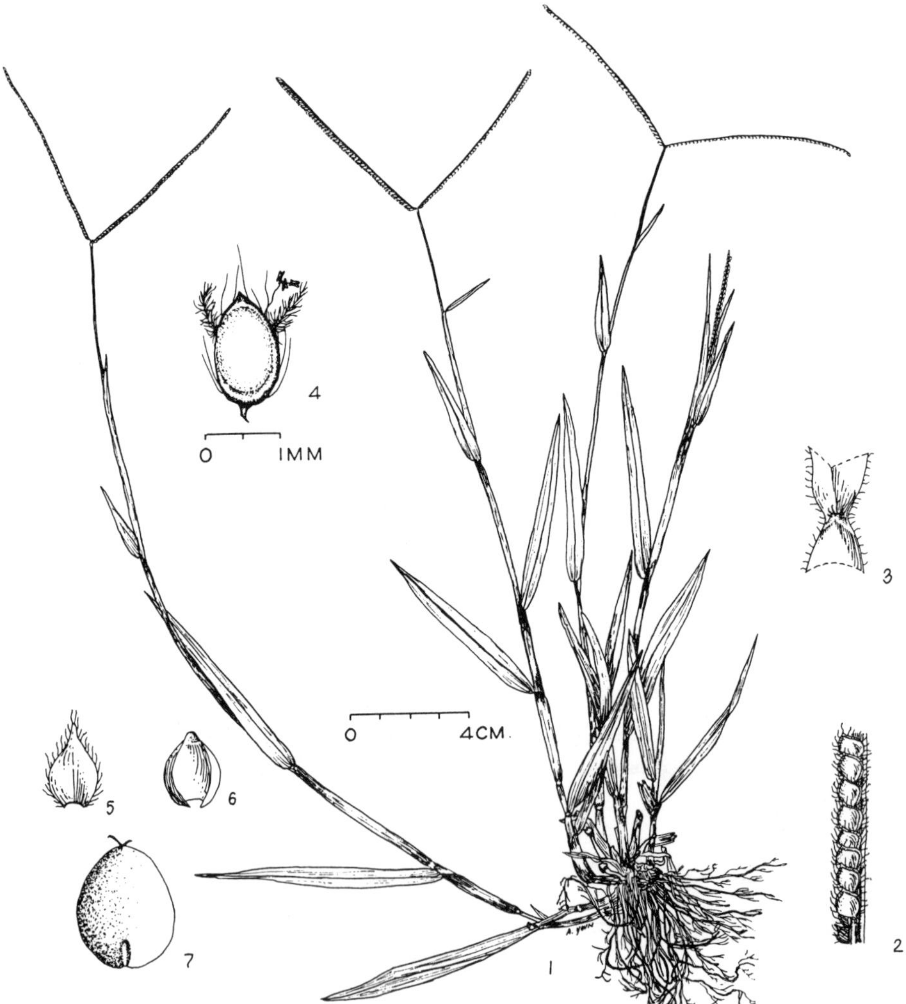

Grassy weeds of importance in the Tropics

Fig. 1.1p *Paspalum conjugatum* Berg. 1. habit; 2. portion of spike; 3. ligule; 4. spikelet; 5. bract; 6. bract; 7. grain (reproduced, with permission of the University Press of Hawaii, from Holm *et al.*, 1977).

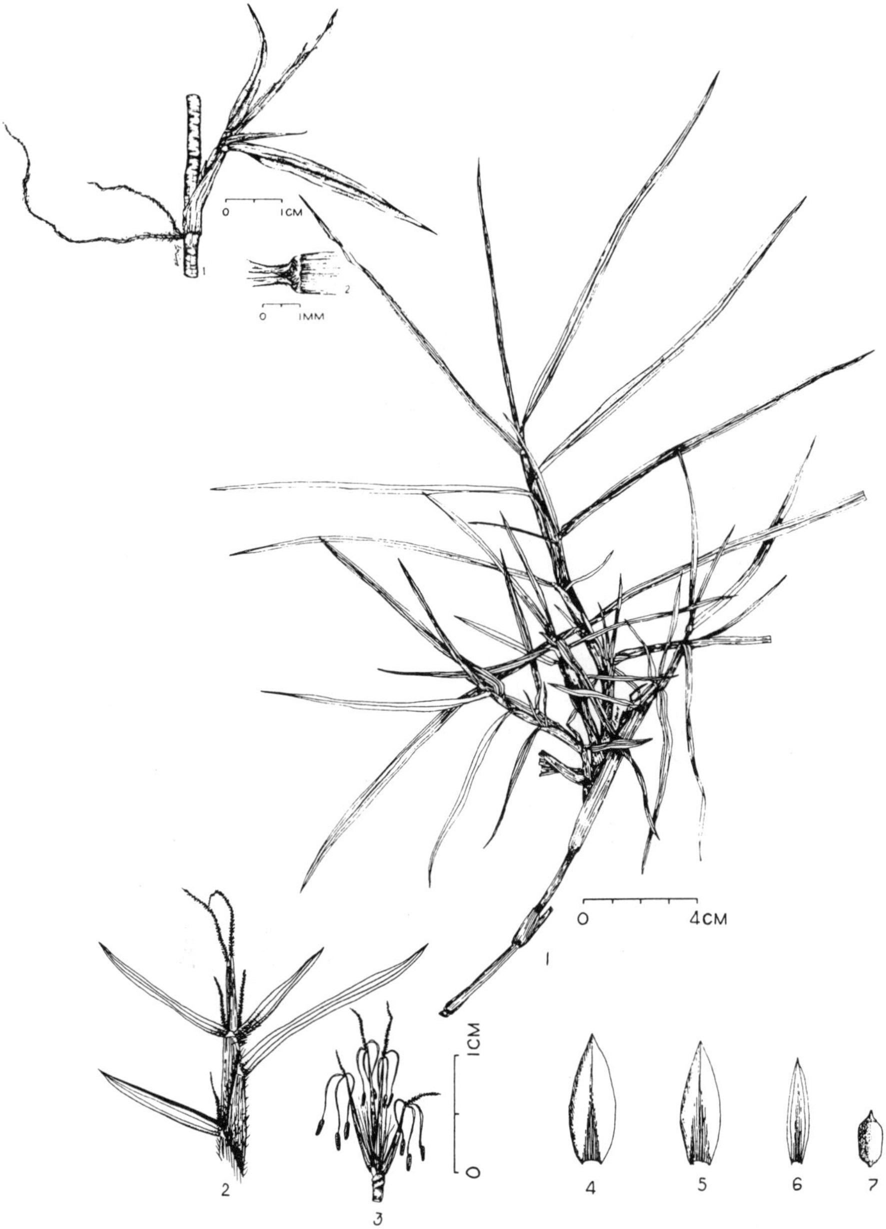

Grassy weeds of importance in the Tropics
Fig. 1.1q *Pennisetum clandestinum* Chiov. 1. habit; 2. flowering culm; 3. spikelet; 4. lower glume; 5. outer glume; 6. lemma; 7. grain. Inset: 1. detail of a young shoot developing from a node of a stolon cutting; 2. ligule (reproduced, with permission of the University Press of Hawaii, from Holm *et al.,* 1977).

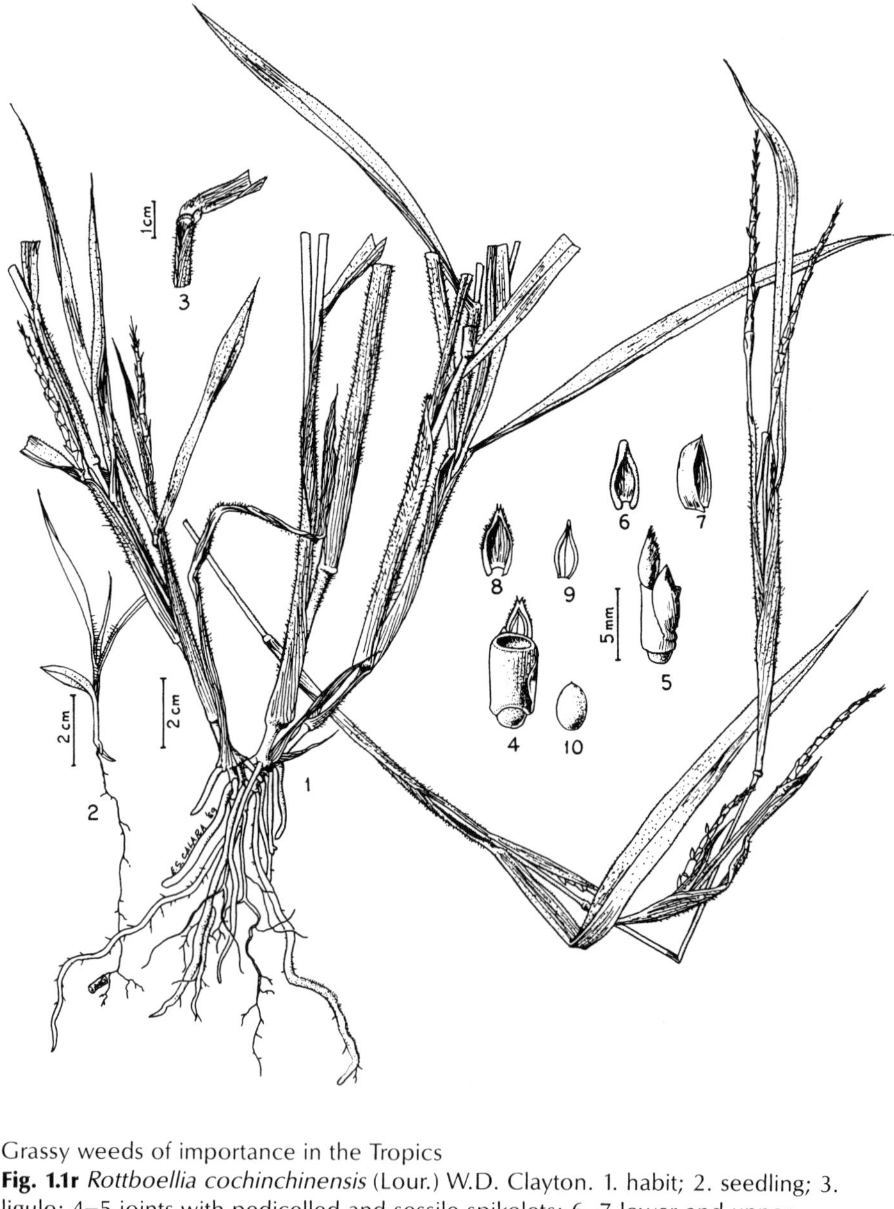

Grassy weeds of importance in the Tropics

Fig. 1.1r *Rottboellia cochinchinensis* (Lour.) W.D. Clayton. 1. habit; 2. seedling; 3. ligule; 4–5 joints with pedicelled and sessile spikelets; 6–7 lower and upper glumes, both of sessile spikelet; 8–9 lower and upper glumes, both of pedicelled spikelet; 10. grain (reproduced, with permission of the University Press of Hawaii, from Holm *et al.*, 1977).

Grassy weeds of importance in the Tropics

Fig. 1.1s *Setaria verticillata* L. P. Beauv. 1. plant habit; 2. spikelets with recurved barbs on the bristles (reproduced, with permission of the University Press of Hawaii, from Holm *et al.*, 1977).

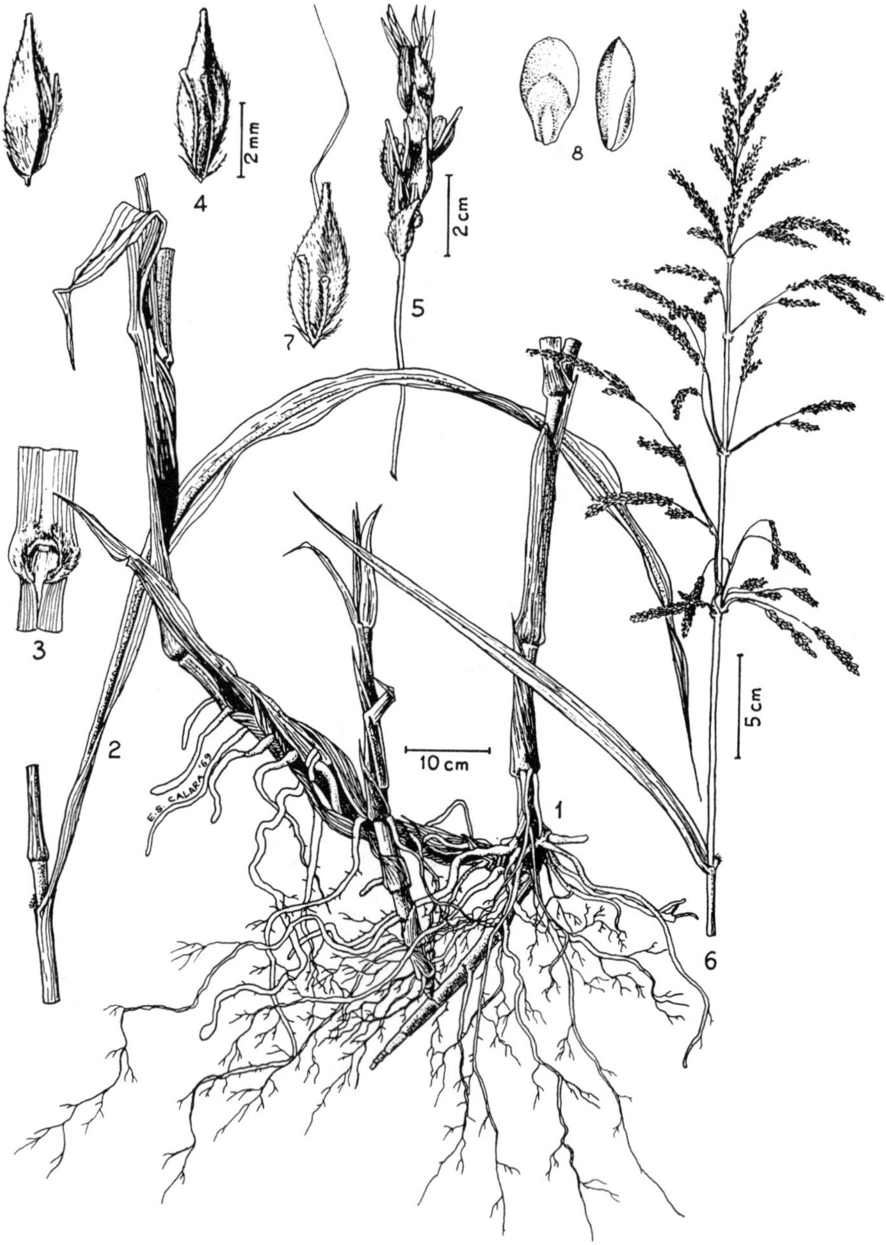

Grassy weeds of importance in the Tropics

Fig. 1.1t *Sorghum halepense* L. Pers. 1. habit; 2. blade; 3. ligule; 4. floret; 5. spikelet; 6. inflorescence; 7. spikelet; 8. grain, two views (reproduced, with permission of the University Press of Hawaii, from Holm *et al.,* 1977).

Chapter 2
The Physiology of Grassy Weeds in Tropical Agriculture

R.H. Groves

CSIRO Division of Plant Industry, GPO Box 1600, Canberra, ACT 2601, Australia

Introduction

The physiology of weedy grasses does not seem to differ greatly from the physiology of non-weedy grasses. The literature on the physiology of grasses is huge, partly because grasses are so common and partly because they are the very basis of human agriculture and grazing systems – indeed, of human well-being. Grasses evolved in the presence of grazing animals and natural fires. Only much later did humans begin to select them for agricultural and specific grazing purposes. Inevitably, however, as agriculture has evolved, undesirable grasses have multiplied and become a nuisance in both temperate and tropical regions.

This chapter concentrates on several aspects of the physiology of grasses which, when integrated with characteristics of the agro-ecosystems they infest, may be associated with weediness in some grasses. These aspects include the physiology of dormancy and germination of grass seeds, the developmental patterns of a few widespread grasses and some characteristics of grasses with different photosynthetic pathways. The review that follows focuses on what is known of the ecophysiology of a few grassy weeds of significance to tropical agriculture, such as *Imperata cylindrica* (blady grass), *Sorghum halepense* (Johnson grass) and *Rottboellia cochinchinensis* (syn. *R. exaltata*) (itch-grass). For some physiological aspects, such as seed dormancy, the information available is much greater for temperate grassy weeds than for tropical species – a situation that may reflect a different level of research effort or, alternatively, a greater physiological and biochemical complexity and a greater ecological significance for dormancy in the seeds of temperate grasses. This chapter is not restricted to only tropical grassy weeds, although the primary emphasis will be on this seemingly intractable group of plants.

Seed dormancy

The process of domestication of cereal grasses has aimed at several physiological features, one of which is the absence of seed dormancy. In this regard Ballard (1964) instanced the extreme uniformity of germination of barley in the malting industry. The opposite situation seems to apply to grassy weeds of both tropical and temperate crops. Fresh seeds of most non-domesticated grasses are physiologically dormant, although morphologically mature. To commence germination, the embryo of seeds of most of these species requires a period of time in which to 'after-ripen', during which biochemical changes occur in the seed that primarily involve qualitative changes in gibberellin composition (e.g. Naylor and Simpson, 1961). In some temperate grassy weeds, as in *Avena fatua*, this after-ripening process is complex and under genetic control.

Whether after-ripening is of much importance in tropical grass species growing in tropical environments is a subject for debate. Fresh seeds of ecotypes of *Sorghum halepense* collected in the United States have a dormant period but this dormancy may be imposed as much by a mechanical restriction of the seed coat as by an after-ripening requirement (Harrington and Crocker, 1923; Taylorson and McWhorter, 1969). Benech Arnold *et al.* (1990a) recently showed that dormancy in fresh seeds of *S. halepense* was overcome by exposure to fluctuating temperatures. Repeated cycles of fluctuating temperatures were additive in their effects, with each cycle resulting in release from dormancy of a further proportion of the population. Such results were used to develop a model of seedling recruitment in the field based on a knowledge of soil temperature regimes (Benech Arnold *et al.*, 1990b). This approach has the potential to unravel the apparent complexities associated with release of seed dormancy in many tropical grassy weeds.

In *Rottboellia cochinchinensis* a variable response to after-ripening has been reported (Holm *et al.*, 1977). For a Kenyan population, Popay (1974) found intact seeds collected from the parent plant to be dormant. Storage of fresh seed at room temperature for five months did not reduce the level of dormancy of intact seeds, but over this period germination of dehusked seeds increased from 16 to 89% – a clear demonstration of after-ripening of the embryo. By eight months from harvest, Popay reported maximal germination. For seed of the same species collected in Zimbabwe, Thomas (1970) reported a similar general result in that the majority of freshly harvested seeds required a period of after-ripening of about four to five months before they could germinate (Table 2.1). In Thomas' case, germination of intact seeds was not promoted by application of either 0.2% KNO_3 or 100 ppm gibberellic acid or by increasing the oxygen supply. Thomas and Allison (1975) subsequently showed that various treatments, such as periodic wetting, dehusking, removal of all bracts, and a combination of wetting and drying in the light, increased germination to various extents. For fresh seeds of *R. cochinchinensis* collected in the Philippines, on the other hand, Moody (pers.comm.) could find no evidence for dormancy, whereas Pamplona and Mercado (1981a) found a variable level of dormancy in five populations from the Philippines. Some of the differences between populations may arise because of different environmental conditions for seed ripening. For instance, Heslop-Harrison (1959a) found that the after-ripening period was briefer for seeds from a genotype of *R. cochinchinensis* collected in northern Australia and grown in short days than for those from plants of the same genotype grown initially in long days.

From what is known of the dormancy responses of fresh seeds of these two grassy weeds, a period of time for physiological after-ripening of the embryo seems to be required, although such a requirement may be modified by morphological restrictions to germination imposed by the bracts.

When populations of the widespread tropical grass *Themeda australis* (syn. *T. triandra*) were collected from both tropical and temperate areas of Australasia and fresh seeds collected from greenhouse-grown plants, seeds of the one population from subtropical Australia (Mundubbera, Queensland) and one from tropical Papua New Guinea (Lae) germinated immediately after collection, compared to seeds from temperate sites which in general needed to after-ripen for a period of three to six months (Groves *et al.*, 1982). The population from tropical Katherine in the Northern Territory, however, responded similarly to the temperate populations (Fig. 2.1). An absence of dormancy in fresh seed in the Lae and Mundubbera populations means that ecologically there is no reason why seed shed in the wet season in the tropics cannot germinate immediately and become established under favourable moisture conditions. There would seemingly be little

Table 2.1. Germination (%) of seeds of *Rottboellia cochinchinensis* with time from harvest and glume removal (after Thomas, 1970).

Seed treatment	*Time from harvest (months)*				
	1	*3*	*4*	*5*	*12*
Intact				0–10	50
Dehusked (glumes removed)	17	37	58	62	95

ecological advantage for the existence of physiological dormancy of fresh seed in such an environment, compared with a temperate situation where germination of *T. australis* seed shed in late summer is optimal in spring when frosts are rare and soil temperatures favourable (Hagon and Groves, 1977). Although *T. australis* is not usually considered weedy, there would be obvious advantages for weedy tropical grasses to show the same adaptation, as seems to be shown by *Imperata cylindrica* for example (Santiago,

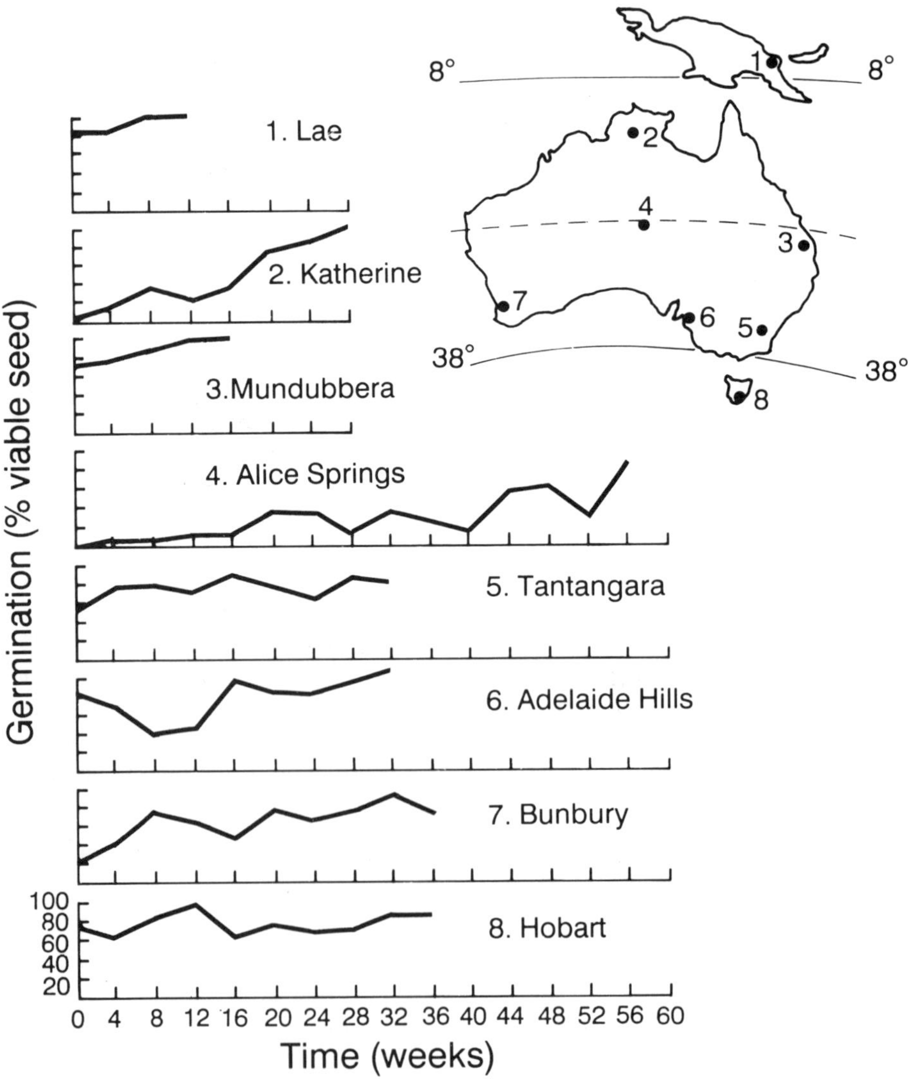

Fig. 2.1 Effect of dry storage on dormancy of different populations of *Themeda australis* (Figure 11.4 of Mott and Groves, 1981).

1965). This conclusion would hold for those grassy weeds in tropical environments with only one wet – dry cycle annually. A greater complexity in dormancy, as shown by *S. halepense* and *R. cochinchinensis*, may be adaptive in those regions such as Kenya where the climatic regime is more complex and involves more than one such cycle annually.

It seems clear that dormancy patterns in tropical weedy grasses are complex and that they vary with population. Any attempts at generalization should not be made until more populations of widespread species are collected and tested under similar conditions and especially in those species, such as *I. cylindrica*, whose distribution spans both temperate and tropical environments.

Germination

Once grass seeds have after-ripened they are then physiologically able to germinate. Seeds of tropical grasses in general germinate maximally at temperatures between 15°C and 35°C. In laboratory tests, alternating temperatures and light may promote germination as Taylorson and McWhorter (1969) found for seeds of *S. halepense* collected from 12 ecotypes and from harvests in each of two years. There is no obvious evidence of variation from this relatively uniform response nor any responses that may differentiate physiology of germination responses of weedy from non-weedy grasses.

Rhizome development

A common feature of grassy weeds of tropical agriculture is their ability to spread vegetatively as well as by seed. In some grassy weeds, such as *Cynodon dactylon*, vegetative spread is the more important means because the species seeds so sparsely. The physiology of rhizome growth and development is thus of considerable importance to a review of the physiology of tropical grassy weeds. Control of rhizome development in grassy weeds in agriculture is essential for effective management. This is especially true of perennial weeds such as *I. cylindrica* and *S. halepense.*

Rhizomes are underground stems that possess buds at intervals along the shoot. Bud development is controlled by temperature (25°C–35°C optimum) and water status, as well as by position of buds along the stem (apical cf. proximal) and a minimal length of stem surrounding the bud. As regards the last factor, more than 10 mm of stem is required in *C. dactylon* (Henson, 1970, as cited by Kasasian, 1971, p. 77). For *I. cylindrica,* Santiago (1965) reported that a piece of rhizome with a minimum of one internode bounded by two nodes was capable of producing a new plant. Soerjani and Soermarwoto (1969) further investigated some factors influencing development and growth of buds in *I. cylindrica*; they found that growth of buds was increased by increased length of rhizome section.

Eussen (1981) investigated in detail the effect of three levels of shading on growth chracteristics of *I. cylindrica.* Rhizomes of shaded plants were shorter and the weight per unit length was less; however, an 80% reduction in light intensity did not influence either rhizome internode length or the number of buds per node on the rhizome (Fig. 2.2). Eussen considered the decrease in weight per unit length caused by heavy shading to be more effective in limiting reproduction than any change in bud number. Further, Eussen noted an irregular occurrence of buds on the rhizomes of *I. cylindrica,* with some rhizomes having a bud at almost every node and others with none at all. Presumably, it is the number of buds towards the apex of the rhizome that are most significant in assessing reproductive potential. Eussen's results have considerable significance to areas being cleared for agriculture as well as to slash-and-burn systems. Weeds such as *I. cylindrica* can halt the succession back to forest in the latter systems, as in northeastern India (Kushwaha and Ramakrishnan, 1982), or render an area

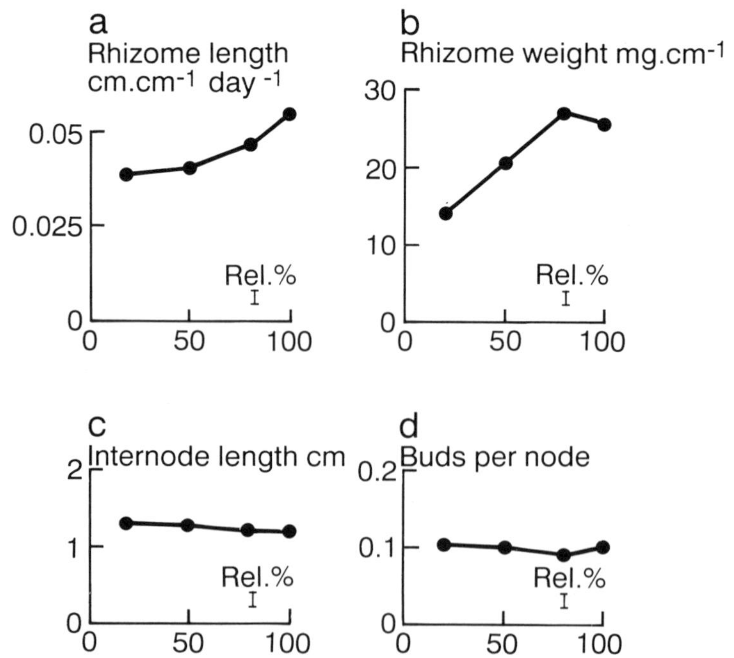

Fig. 2.2 Effect of decreased light intensity (20%, 50% or 80% of full sunlight) on (a) length, (b) dry weight per cm, (c) internode length and (d) number of buds per node, for rhizomes of *Imperata cylindrica* (Figures 6 and 7 of Eussen, 1981).

unsuitable for agriculture, as in Indonesia (Eussen, 1981).

Growth of *I. cylindrica* plants arising from rhizomes collected in the southern United States was higher at a temperature regime of 29°C (day)/ 23°C (night) than at 23°C (day)/17°C (night) and higher at 16 h daylength than at 12 h (Patterson *et al.*, 1980). The numbers and weights of rhizomes were also greater at the higher temperature and longer daylength. However, the partitioning of total plant weight into rhizomes was greater at the lower temperature (but at the longer daylength) for plants started from rhizomes. Perhaps more significantly in the context of this review, there were differences between populations in some growth characteristics – this will be discussed later.

Kushwaha *et al.* (1983) studied a slash-and-burn agricultural system in northeastern India in which *I. cylindrica* is a major weed. They showed that plants arising from rhizomes were more competitive in mixtures than plants arising from seeds, at least in the early stages of growth. Presumably, this difference arose from the larger 'capital' available for regrowth from rhizomes than that in the small seeds. Kushwaha and Ramakrishnan (1982) showed that the allocation of biomass to underground organs of *I. cylindrica* was higher in plants grown in sunlight compared with those grown in shade, thereby suggesting a greater potential for vegetative reproduction in sunlight than in shade. More results on the physiology of rhizome bud development in grassy weeds may add to the effectiveness of control measures against these major perennial weeds which are not only so prevalent in plantation crops but also in slash-and-burn agricultural systems.

Growth

Most of the grasses that are weedy in tropical agro-ecosystems initially fix carbon as four-carbon acids – the so-called C4 pathway of photosynthesis. Possession of this C4 pathway confers some advantages for plant growth in terms of photosynthetic efficiency under conditions of high light intensity, high temperatures and limited moisture availability – conditions common in tropical environments. Some of the characteristics of the C3 (Calvin cycle) and C4 pathways for CO_2 fixation are summarized in Table 2.2. Whilst the subdivision of plants into C3 and C4 is of special interest to biochemists and plant physiologists, it has aroused considerable interest amongst weed scientists since Holm *et al.* (1977) highlighted the prevalence of the C4 pathway in the world's worst weeds. The C4 grass weeds of tropical systems include *Cynodon dactylon, Imperata cylindrica, Sorghum halepense, Rottboellia cochinchinensis, Eleusine indica, Echinochloa* spp. and *Paspalum conjugatum*, as well as the sedges *Cyperus rotundus* and *C. esculentus.*

Table 2.2. Comparison of photosynthetic pathways (abridged table 5.1 from Radosevich and Holt, 1984).

Characteristic	C_3	C_4
Leaf anatomy	Palisade + spongy parenchyma	Non mesophyll differentiation; large bundle sheath; Kranz-type
Light saturation point	6,000 ft-c	8,000-10,000 + ft-c
Optimum temperature (°C)	20–30	30–45
Maximum photosynthetic rate (mg/dm^2/hr)	30	60
Maximum photosynthetic rate (mg/g/hr)	55	100
Maximum growth rate (g/dm^2/day)	1	4
Water use efficiency (g CO_2/g H_2O)	300	600
Photorespiration	High	Low
Na required	No	Yes
Fixation path and enzyme	CO_2 + 5-C → 3-C PGA; carboxydismutase or also called ribulose bisphosphate carboxylase	CO_2 + 3-C → 4-C acids; PEP carboxylase
Space–time relations	Entire Calvin cycle in any mesophyll cell	Initial fixation in mesophyll, then transfer of acid to bundle sheath for Calvin cycle
CO_2 compensation point	50 ppm	5 ppm

Vegetative growth of most C4 grassy weeds is limited at day temperatures below about 20°C and is maximal at day temperatures between 30°C and 35°C. Night temperatures below 15–20°C strongly limit growth. For instance, Patterson and Flint (1979) and Patterson *et al.* (1979) found growth of *R. cochinchinensis* to be optimal at 32°C (day)/26°C (night) and to be greatly retarded at day temperatures below 29°C and night temperatures below 23°C.

Despite an increasing range of literature on the physiology of C4 grasses, differences in the factors affecting vegetative growth of weedy and non-weedy tropical grasses are not obvious as yet.

Flowering

Most tropical grasses appear to require short days (12 h or less) to initiate floral primordia at the apex. The expression of those primordia as flowers may be delayed, however, until the onset of the wet season which may occur in longer daylengths. Evans (1964) reviewed what information was available on flowering responses among the different taxonomic groups of grasses; in the Panicoideae, which includes most of the tropical grasses, there was a predominance of short-day responses and no confirmed cases of an obligate long-day response in most grass species in the Chloridoideae (Table 8.1 in Evans, 1964). The only confirmed long-day response for floral initiation was in a population of *Cynodon dactylon* in the Chlorideae presumably collected at relatively high latitudes in the vicinity of Pretoria, South Africa by Mes (see Evans, 1964, for reference).

In one widespread tropical grass, differences between populations in terms of flowering response have been reported. Evans and Knox (1969) found that populations of *Themeda australis* collected from New Guinea and northern tropical Australia usually required short days for floral initiation whilst populations from southern Australia usually required long days. Populations from intermediate latitudes were mainly day neutral. Populations of *T. australis* from southern Africa (South Africa and Zimbabwe) were all long-day plants (Downing and Groves, unpublished). An approach similar to that of Evans and Knox (1969) but using populations of *Imperata cylindrica* collected over an even wider latitudinal transect may be especially revealing. Heslop-Harrison (1959b) found Australian material of *Rottboellia cochinchinensis* collected from a tropical site to give a short-day response for flowering. Evans (1962) reported that in plants of the same genotype, inflorescence initiation occurred in daylengths up to 11 h but none in plants given 13+ h. Pamplona and Mercado (1981b) grew five populations of *R. cochinchinensis* from the Philippines and found considerable variation in the critical daylength for flowering, ranging from just under 12 h in one population to 14 h in another. Evans (1989) showed the existence of photoperiodically sensitive and insensitive lines in the one population of *R. cochinchinensis* collected from Lae, Papua New Guinea. Genotypes that are relatively insensitive to daylength, such as found in Papua New Guinea and the Phillipines, may pose serious weed problems in crops grown at higher latitudes; in fact, Evans (1989) claims that such genotypes may already exist in the United States, because Patterson *et al.* (1979) recorded rapid flowering in *R. cochinchinensis* plants grown in 16 h days even at the relatively low day/night temperatures of 26°C/14°C, 20°C/20°C and 17°C/23°C. Given such apparent variation in flowering response between populations, such as that found by Pamplona and Mercado (1981b) in the Philippines, experimentation with populations collected even more widely seem required in the cases of *R. cochinchinensis* and also of *I. cylindrica*. The claim of greater weed potential of photoperiodically insensitive genotypes in *R. cochinchinensis* in areas of higher latitude by Evans (1989) seems especially significant in this regard.

Conclusions

Are there any physiological responses typical of tropical grassy weeds as a group? Grassy weeds, e.g. *I. cylindrica* and *S. halepense*, tend to be perennial, and nuisances in plantation crops and slash-and-burn agricultural systems. But some major weeds, such as *R. cochinchinensis* and *Echinochloa* spp., are annuals and are weedy in land that is cropped annually. Tropical grass weeds reproduce usually both as seeds and from rhizomes, their seeds are often initially dormant, the plants grow maximally at temperatures between 25°C and 35°C at high light intensities and do not grow at temperatures below about 18°C, they possess C_4 photosynthetic characteristics and require short days to initiate their flowers. Exceptions to this 'ideotype' exist, however, at both the species and population level, as shown earlier. In reviewing both seed dormancy and flowering of tropical grasses, it has been noted that different populations may behave differently and that collections of material from a wide range of different environments may be necessary to show whether a characteristic and general physiological response exists among tropical grasses. Such a collection has been the case only for *Themeda australis*, and as far as the author can determine, never for weedy tropical grasses. Just as many of the tropical weeds have wide distributions, they are also very variable morphologically. For instance, Hubbard (1944) described five varieties of *I. cylindrica*, each of which is distributed over a definite geographical region. Within some of his taxonomically delineated varieties Hubbard further identified a series of habitat forms. *Sorghum halepense* is another variable grass species. The first conclusion that may be drawn is that the existence of variation, both physiological and morphological, rules out most generalizations about physiological responses of tropical grasses and not just the weedy species.

It can be concluded that just as there is no one tropical climate there is no general physiological response to the variables associated with tropical environments. With the apparent exception only of *Cynodon dactylon* (Evans, 1964), all other tropical grass species seem to require short days for floral initiation. A further generalization is that agamospermous apomixis is the predominant type of reproduction in a large number of grasses from tropical and subtropical regions. Widespread tropical grasses such as *Heteropogon contortus*, *Themeda triandra* and *Bothriochloa ischaemum* are all facultatively apomictic (Brown and Emery, 1958; Emery and Brown, 1958; Evans and Knox, 1969). Even these generalizations do not apply to populations of some tropical grasses growing in temperate high latitude environments, such as *T. triandra* in southern Australia and southern Africa. Nor do they necessarily apply to tropical grasses such as *Dichanthium aristatum* of Australian origin which normally needs short days to flower but in which the degree of apomixis is less in long days, i.e. a higher level of sexual reproduction (Knox and Heslop-Harrison, 1963). Thus the light regime in which tropical grasses are grown may influence the balance between apomixis and sexuality.

Whilst it is difficult to conclude that there are physiological attributes shared among tropical grasses it is even more difficult to identify any physiological attributes predominant in weedy tropical grasses. As with Baker's 'ideal weed' (Baker 1965), so with the subset of weedy tropical grasses; there is no one attribute – physiological or otherwise – that is associated with weediness. It is more likely to be a combination of attributes which in the case of weedy tropical grasses include the highly plastic responses to environmental factors in relation to seed dormancy and flowering in particular, together with the existence of intraspecific variation, plasticity in breeding systems and the possession of certain morphological attributes, such as rhizome development. By learning more of the physiology of these attributes, the task for controlling tropical grassy weeds may become less intractable in the future.

Acknowledgement

I thank my colleague Lloyd Evans for constructive comments on a first draft of this chapter.

References

Baker, H.G. (1965) Characteristics and modes of origin of weeds. In: Baker H.G. and Stebbins G.L. (eds) *The Genetics of Colonizing Species.* Academic Press, New York, pp147–68.

Ballard, L.A.T. (1964) Germination. In: Barnard C. (ed.) *Grasses and Grasslands.* Macmillan, London, pp73–88.

Benech Arnold, R.L., Ghensa, C.M., Sanchez, R.A. and Iusaushi, P. (1990a) Temperature effects on dormancy release and germination rate in *Sorghum halepense* (L.) Pers. seeds: a quantitative analysis. *Weed Research* 30, 81–9.

Benech Arnold, R.L., Ghensa, C.M., Sanchez, R.A. and Iusaushi, P. (1990b) A mathematical model to predict *Sorghum halepense* (L.) Pers. seedling emergence in relation to soil temperature. *Weed Research* 30, 91–9.

Brown, W.V. and Emery, W.H.P. (1958) Apomixis in the Gramineae: Panicoideae. *American Journal of Botany* 45, 253–63.

Emery, W.H.P. and Brown, W.V. (1958) Apomixis in the Gramineae tribe Andropogoneae: *Heteropogon contortus. Madrono* 14, 238–46.

Eussen, J.H.H. (1981) Studies on the growth characteristics of alang-alang (*Imperata cylindrica* (L.) Beauv.). *BIOTROP Bulletin* 18, 1–23.

Evans, L.T. (1962) Day-length control of inflorescence initiation in the grass *Rottboellia exaltata* L. f. *Australian Journal of Biological Sciences* 15, 291–303.

Evans, L.T. (1964) Reproduction. In: Barnard, C. (ed.) *Grasses and Grasslands.* Macmillan, London, pp126–53.

Evans, L.T. (1989) *Rottboellia cochinchinensis.* In: Halevy, A.H. (ed.) *CRC Handbook of Flowering, Vol. VI.* CRC Press, Boca Raton, Florida, USA, pp547–52.

Evans, L.T. and Knox, R.B. (1969) Environmental control of reproduction in *Themada australis. Australian Journal of Botany* 17, 375–89.

Groves, R.H., Hagon, M.W. and Ramakrishnan, P.S. (1982) Dormancy and germination of seed of eight populations of *Themeda australis. Australian Journal of Botany* 30, 373–86.

Hagon, M.W. and Groves, R.H. (1977) Some factors affecting establishment of four native grasses. *Australian Journal of Experimental Agriculture and Animal Husbandry* 17, 90–6.

Harrington, G.T. and Crocker, W. (1923) Structure, physical characteristics and composition of the pericarp and integument of Johnson grass seed in relation to its physiology. *Journal of Agricultural Research* 23, 193–222.

Heslop-Harrison, J. (1959a) Photoperiod effects on sexuality, breeding system and seed germinability in *Rottboellia exaltata. Proceedings of the IX International Botanical Congress, Montreal, Vol. 2.* University of Toronto Press, Toronto, pp162–3.

Heslop-Harrison, J. (1959b) Photoperiod and fertility in *Rottboellia exaltata* L. f. *Annals of Botany (N.S.)* 23, 345–9.

Holm, L.G., Plucknett, D.L., Pancho, J.V. and Herberger, J.P. (1977) *The World's Worst Weeds. Distribution and Biology.* The University Press of Hawaii, Honolulu, 609pp.

Hubbard, C.E. (1944) Taxonomy, description and distribution of species and varieties. In: *Imperata cylindrica. Taxonomy, Distribution, Economic Significance and Control.* Imperial Agricultural Bureaux Joint Publication No. 7, Imperial Bureau of Pastures & Forage Crops, Aberystwyth, Great Britain, pp5–13.

Kasasian, L. (1971) *Weed Control in the Tropics.* Leonard Hill, London.

Knox, R.B. and Heslop-Harrison, J. (1963) Experimental control of aposporous apomixis in a grass of the Andropogonae. *Botaniska Notiser* 116, 127–41.

Kushwaha, S.P.S. and Ramakrishnan, P.S. (1982) Observations on growth of *Eupatorium odoratum* L. and *Imperata cylindrica* (L.) Beauv. var. *major* under different light and moisture regimes. *Proceedings of the Indian National Science Academy* B48, 689–93.

Kushwaha, S.P.S., Ramakrishnan, P.S. and Tripathi, R.S. (1983) Competitive relationships of the plants of *Imperata cylindrica* established from rhizomes and from seeds. *Tropical Plant Science Research* 1, 53–7.

Mott, J.J. and Groves, R.H. (1981) Germination strategies. In: Pate, J.S. and McComb, A.J. (eds) *The Biology of Australian Plants.* University of Western Australia Press, Nedlands, pp307–41.

Naylor, J.M. and Simpson, G.M. (1961) Dormancy studies in seed of *Avena fatua.* 2. A gibberellin-sensitive inhibitory mechanism in the embryo. *Canadian Journal of Botany* 39, 281–95.

Pamplona, P.P. and Mercado, B.L. (1981a) Ecotypes of *Rottboellia exaltata* L.f. in the Philippines. I. Characteristics and dormancy of seeds. *The Philippine Agriculturist* 64, 59–66.

Pamplona, P.P. and Mercado, B.L. (1981b) Ecotypes of *Rottboellia exaltata* L.f. in the Philippines. II. Response to daylength and nitrogen application. *The Philippine Agriculturist* 64, 371–8.

Patterson, D.T. and Flint, E.P. (1979) Effects of simulated field temperatures and chilling on itch grass (*Rottboellia exaltata*), corn (*Zea mays*), and soybean (*Glycine max*). *Weed Science* 27, 645–50.

Patterson, D.T., Flint, E.P. and Dickens, R. (1980) Effects of temperature, photoperiod, and population source on the growth of cogongrass (*Imperata cylindrica*). *Weed Science* 28, 505–9.

Patterson, D.T., Meyer, C.R., Flint, E.P. and Quimby, P.C. Jr. (1979) Temperature responses and potential distribution of itch-grass (*Rottboellia exaltata*) in the United States. *Weed Science* 27, 77–82.

Popay, A.I. (1974) Investigations into the behaviour of the seeds of some tropical weeds. 1. Laboratory germination tests. *East African Agricultural and Forestry Journal* 40, 31–43.

Radosevich, S.R. and Holt, J.S. (1984) *Weed Ecology. Implications for Vegetation Management.* John Wiley & Sons, New York.

Santiago, A. (1965) Studies in the autecology of *Imperata cylindrica* (L.) Beauv. (1812). *Proceedings of the 9th International Grasslands Congress, Sao Paulo,* 499–502.

Soerjani, M. and Soermarwoto, O. (1969) Some factors affecting germination of alang-alang (*Imperata cylindrica*) rhizome buds. *Pest Articles and News Summaries* 15, 376–80.

Taylorson, R.B. and McWhorter, C.G. (1969) Seed dormancy and germination in ecotypes of johnsongrass. *Weed Science* 17, 359–61.

Thomas, P.E.L. (1970) A study of the biology of *Rottboellia exaltata* Linn. f. *Proceedings of the 10th British Weed Control Conference,* 669–76.

Thomas, P.E.L. and Allison, J.C.S. (1975) Seed dormancy and germination in *Rottboellia exaltata. Journal of Agricultural Science, Cambridge* 85, 129–34.

Chapter 3
Biological Control of Tropical Grassy Weeds

H.C. Evans

International Institute of Biological Control
(IIBC, An Institute of CAB International),
Silwood Park, Buckhurst Road, Ascot,
Berks, SL5 7TA, UK

General introduction

Research on the biological control of grassy weeds is of recent origin and, not surprisingly, there is a relative dearth of information on the subject; most of it relating to temperate ecosystems. The aim of this chapter, therefore, is to briefly present examples of successful biological control projects against broadleaved weeds in temperate crops, in order to demonstrate the principles involved, and then to review a current biological control project relating to a tropical grassy weed, and finally to overview the potential of natural enemies to control the major tropical grassy weeds. Since this chapter discusses weed–natural enemy interactions involving only fungal pathogens, it may appear that, as the author is a mycological pathologist, this chapter is somewhat biased in its choice of examples. However, for once entomological colleagues are in agreement when it is concluded, at this early stage in the proceedings, that the prospects of using arthropod agents to control tropical grassy weeds are remote. Indeed, entomologists have consistently failed to offer any past or present examples of the employment of arthropods to control grassy weeds. It would be appropriate here to quote relevant statements as to why this should be, nevertheless, it seems that none have been made in the biological control literature. Graminaceous weeds are not included in the world catalogue of biological control agents and their target weeds (Julien, 1982; 1987), which deals predominantly with arthropod natural enemies. The over-riding reason for this apparent omission would seem to be general absence of host specific insects associated with grassy weeds. Simmonds (1972), for example, refers to projects being carried out by IIBC on tropical weeds: 'We now have, after a number of years of research, some very promising, highly host specific natural enemies' (in those days only arthropod agents were considered) 'including *Imperata cylindrica* and *Cyperus rotundus*'. Despite this optimism, none were found to be sufficiently damaging or specific to warrant further investigation. One of these surveys was reported by Girling (1972) who failed to collect any suitable monophagous insect agents on *Imperata cylindrica* in East Africa. Holm *et al.* (1977) record the presence of polyphagous insects on *I. cylindrica* which also attack cultivated cereals. Similarly, many of the sap-sucking insects feeding on grass hosts are non-selective and are thereby vectors of numerous virus diseases which they transmit from graminaceous crop plant to grassy weed and vice versa. The diversity of polyphagy is directly linked to plant architecture (Strong *et al.*, 1984). Thus, the simplistic form of the Gramineae–unbranched, strap-shaped leaves (common to both grasses and sedges) reduces the evolution of monophagy and promotes polyphagy. Indeed, Strong *et al.* (1984) exclude grasses from consideration when reviewing the subject of phytophagous insect diversity, because none of the food plant records are reliable within the Gramineae. Moore (pers.comm.) argues that certain insects which attack apical meristems, such as stem borers, damage plants far out of proportion to the biomass consumed and can have acceptably high host specificity (Nye, 1958). Such insects can also directly affect the species composition of grassy ecosystems (East *et al.*, 1986) and hence could be potentially useful biological control agents of grassy weeds. However, because this is still largely theoretical, all the case studies presented in this paper are fungal pathogen-orientated. As this is a specialized research area and the symposium is for generalists, it is necessary to outline the concepts of biological control in relation to tropical grassy weeds before proceeding further.

Concepts

What is biological control?

Biological control is the use of one or more organisms (biotic agents) to maintain another organism (the pest) at a level at which it is no longer a problem: potentially exploiting any organism which interacts with the target pest. Originally, the term 'biological control' was applied to the transfer of a natural enemy of the pest from one area (the centre of origin or co-evolution) to another where it does not occur with the objective of controlling it in the new (exotic) area. This is classical biological control and ideally results in a self-sustaining balanced system which maintains the pest population at a non-damaging level. Classical biological control operates relatively slowly and may not reduce the pest to below the desired economic threshold, thus it may be necessary to improve control by periodic releases of mass-produced agents. This is augmentative or inundative biological control and can involve both indigenous and exotic agents. Typically, biological control is only considered as a last resort when other control measures have failed or are uneconomic, or, as of more recent times, adjudged to be environmentally undesirable. This recent awareness of the often detrimental ecological impact of modern farming techniques has led to a third form of biological control, integrated pest management, in which indigenous natural enemies are conserved, and possibly enhanced, by manipulation of one or more components in the agro-ecosystem (for example, modifying the input of chemicals).

How does it work for weeds?

The aim of the classical strategy is merely to reduce the vigour of the weed to below the 'pest threshold', not usually to kill or eliminate it. Generally, agents are selected on the basis of their specificity and ability to inflict significant damage on the target weed. Biological control of weeds has traditionally been implemented by releasing phytophagous arthropods which damage the photosynthetic and/or reproductive capacity of the pest plant. It is a science of long-standing, dating back to the 19th Century (Julien, 1982; Schroeder, 1983). Conversely, the use of microbial agents or pathogens for weed control has only recently been examined and the potential of many weed–pathogen associations remains to be evaluated. Efforts to control weeds biologically with plant pathogens have consequently lagged behind those of entomologists but, since 1970, there has been a rapid increase in interest in the exploitation of pathogens, particularly fungi, for weed control (Freeman, 1981). With classical biological control, it is often considered that weeds of permanent and semi-permanent situations (forests, pastures) are much better targets than those of disturbed agro-ecosystems (arable crops). However, there are notable exceptions and these will be discussed later. The alternative biological control strategy for weeds involves the bioherbicidal approach in which the agent, invariably a fungal pathogen, is applied to the target weed using the same tactics as for chemical herbicides. The formulated agent or product, termed a mycoherbicide, is sprayed onto the weed to uniformly kill or suppress growth. Formulation increases the pathogenicity, viability and dispersability of the fungal agent, overcoming natural constraints which reduce the effectiveness of the pathogen under normal field conditions.

How is it implemented?

Searches are made for natural enemies of the target weed, both in the literature (including herbarium and museum records) and in the field. However, there is frequently little information available on the natural enemies of non-crop plants (weeds) and much of it is in the form of unpublished reports and collection notes. For the latter, we owe much to the scholarship, industriousness and forward-thinking of a handful of colonial pathologists and entomologists who had the foresight – so lacking today where short-term profit and job security are the driving forces – of the true scientific collector: to gather and collate obscure organisms on seemingly insignificant and economically unimportant plant hosts.

Two field surveys are required: one in the target area and one in the indigenous range of the weed (Schroeder, 1983; Cock, 1986). These may, of course, coincide as in the case of searches for agents with biopesticide ability (the inundative strategy). Thus, the initial surveys are undertaken in the target area in order to determine if any of the natural enemies from the centre of origin of the weed are already present: essential to avoid possible waste of resources, not to mention embarrassment. Once these have been completed, exploratory surveys are conducted in the indigenous range of the weed. There are abundant data to show that the diversity of oligophagous and monophagous natural enemies will coincide with the region of origin and the centre of diversity of the genus of a weed (Schroeder, 1983; Schroeder and Goden, 1986). If these are unknown, and often the botanists have conflicting views, then a wider ranging survey is necessary to try and locate the area with the greatest diversity of natural enemies associated with the weed. The biocontrol specialist may then be in a position to inform the botanist of the co-evolutionary centre of that particular plant species or genus: 'Thus, as mycology matures, it may well be able to give more aid to phanerogamic botany than the science of phanerogamic botany formerly gave to mycology' (Savile, 1954).

Once potential agents have been provisionally assessed in the field for specificity and ability to cause significant damage to the target weed, there then follows a laboratory and greenhouse-based phase to fully determine the taxonomy, biology and specificity of the selected agent. This can often be a prolonged process, which contrasts sharply with the early, ill-conceived (non-specific) attempts at biological control which are frequently resuscitated and quoted by detractors of this approach to pest control (Simmonds, 1972). Simmonds and others (Julien, 1982; Schroeder, 1980; Schroeder and Goeden, 1986; Harris, 1988) point out that, since the scientific basis of biological control was established, many thousands of natural enemies have been moved around the world, or otherwise manipulated, without any detrimental environmental effects or other mishaps. Indeed, rigid guidelines have been laid down for evaluating the safety and specificity of organisms for the biological control of weeds. These comprise a centrifugal phylogenetic testing method, involving the exposure to the potential biocontrol agent of a sequence of plants from those most closely related to the weed to successively more and more distantly related plants, and an in-built safeguard, in which a range of well-defined criteria are used to select additional test species (Wapshere, 1974). The same author has recently re-evaluated the screening procedures and now considers them to be too strict: this could lead to the rejection of potentially useful organisms due to over-zealous interpretation of greenhouse test results (Wapshere, 1989).

How does it differ in the tropics?

Simmonds (1972) considered that, theoretically, there are no differences in undertaking biological control in the tropics compared with temperate climates: 'In practice, however, there may be considerable differences'. One immediate difference is that in many tropical zones, the pests and their natural enemies breed quickly and continuously and thus biological control, if successful, is both rapid and spectacular. Whilst the more cynical view of Cock (1986) is that there are gross disparities between biological control in the tropics and that elsewhere, which is a reflection of the wealth of the countries concerned. Hence, although numerous introductions of natural enemies have been made in the tropics, a high proportion of the projects have been inadequately funded and, as a consequence, have failed due to poor management and/or lack of continuity.

Why employ it to control tropical grassy weeds?

Because vast areas of the tropics are underdeveloped and their general economy is at a low level, chemical control methods, even when these are of proven economic value, may not be extensively used or applications very carefully carried out for lack of expertise and hence have their limitations. I would, therefore, suggest that far greater attention be paid to biological control in the tropics ...

(Simmonds, 1972)

The author was referring to classical biological control only and not specifically to grassy weeds, but, with the increasing cultivation and importance of graminaceous crops in tropical Africa – Niger, 4.5 million ha; Nigeria, 3.4 million ha; Mali, 1.4 million ha (Anon., 1989) – grassy weeds will present an ever increasing and major control problem due to lack of specificity of chemical herbicides. If we examine the case of wheat in temperate countries, heavy reliance has been placed on chemical weed control

(Lisansky, 1989). The direct cost of weed control, plus losses due to weeds, is estimated at US$5 billion in the USA alone, although future sales may be adversely affected due to the evolution of weed resistance to chemical herbicides. Lisansky quotes the example of chlorsulfuron herbicide which has already replaced 40% of the herbicide market for winter wheat in the USA. This compound has properties that can cause rapid development of resistance to it as well as to other herbicides. Ultimately, this will result in a loss of confidence by farmers as they discover that previously reliable products no longer work. The weeds most difficult to control in wheat belong to related graminaceous genera: *Alopecurus, Apera, Avena, Lolium* and *Setaria*, which become highly competitive in the absence of broad-leaved weeds and difficult to control due to their morphological, physiological and biochemical similarities with the crop. They can significantly reduce yield at low densities and some can suppress yield irreversibly, suggesting allelopathic interactions. In addition, grassy weeds often respond more vigorously to fertilizers than the crop, thereby increasing their competitiveness.

Temperate grassy weeds (*Alopecurus myosuroides, Avena fatua, Lolium rigidum*) have evolved resistance to a variety of chemical herbicides, even to selective ones which were employed previously to control them in wheat. The situation, therefore, is becoming extremely complex and often a cocktail of chemical herbicides may be required because those active against broadleaved weeds do not control grasses and most active graminaceous herbicides control few broadleaved weeds and usually only one grassy weed. 'Anything that can be agronomically and economically done to cost-effectively sustain yields and delay herbicide resistance is worthwhile' (Lisansky, 1989). If this is the plea now for temperate agriculture, what will be the situation in the tropics in 10–20 years following the projected donor-aided expansion of graminaceous crops? The time would seem to be right to examine the potential of biological control for the long term (sustained) control of tropical grassy weeds. To quote Holm (in Worsham, 1982):

> Perhaps what this discipline needs is to strike out boldly for the biological control of *Eichhornia crassipes, Cyperus rotundus*, the world's No. 1 weed, *Imperata cylindrica, Echinochloa crus-galli*, the worst weed of rice, the world's largest crop, *Hydrilla verticillata, Sorghum halepense* and a host of other important weeds. Through these efforts more children will have food, costs will be reduced and the environment will be spared large amounts of chemicals.

In this paper, the evidence for and against biological control of tropical grassy weeds will be examined in less emotive terms, and recommendations made for future action. It is significant, perhaps, that the weed examples listed by Holm are predominantly tropical grassy species; the sedges are included within this general grouping since they have many characteristics in common with the Gramineae.

CASE STUDIES

Broadleaved weeds

Classical biological control

As previously mentioned, rust fungi are extremely efficient classical biological agents of weeds; consequently they have been the group of pathogens most investigated. One of the pioneering and still most successful examples of this form of control involves the skeleton weed, *Chondrilla juncea* (Compositae), and the rust, *Puccinia chondrillina* Bubak & Sydow in Australia. The plant is of Mediterranean origin which, after its accidental introduction, became a major weed of wheat and fallow land in areas of southeast Australia with a Mediterranean-type climate. Severe rust infection results in either host death or physiological stress, with subsequent reduction in seed production and storage reserves. After pathogenicity and specificity tests in Europe, lasting three years and involving 56 plant species in 33 families (Hasan and Wapshere, 1973), the rust was introduced into southeast Australia where it spread rapidly, covering hundreds of kilometres in a few months (Cullen *et al.*, 1973). Weed infestations have been reduced by 99%, to a density approaching that normally observed in Europe (Cullen and Hasan, 1988), with an estimated saving of Aus$16 million per annum from increased yields and reduced herbicide costs (Mortensen, 1986). 'Compare this figure with the cost of the original project (a gross overestimate would probably be half a million Australian dollars) and the financial, to say nothing of the ecological benefits of classical biological control, become evident' (Evans, 1987a). A number of similar projects are currently underway involving exotic weeds which are difficult or uneconomic to control chemically. IIBC has several tropically orientated projects employing rusts against various broadleaved weeds: *Parthenium hysterophorus* (Compositae), *Cryptostegia grandiflora* (Asclepiadaceae) and *Chromolaena odorata* (Compositae) (Evans, 1987a and b).

Inundative biological control (mycoherbicides)

There is increasing optimism concerning the use of mycoherbicides to control agricultural weeds, broadleaved species at least. Several products are now on the market and many more are in the research pipeline (Cullen and Hasan, 1988). The pioneering technology was developed in the USA over a number of years, and North America, in general, is still the focus of mycoherbicide expertise. In particular, species of the genus *Colletotrichum* are attracting attention because of their aggressiveness (often causing lethal stem cankers), ease of cultivation and high specificity: physiological races, pathotypes or *formae speciales* have been identified in a number of taxa. The history surrounding the development of the first product, Collego, based on *Colletotrichum gloeosporioides* (Penz.) Sacc. f. sp. *aeschynomene*, for control of northern jointvetch (*Aeschynomene virginica*) in rice and soyabean in southern USA, has been well documented (Templeton *et al.*, 1979; Templeton and Greaves, 1984; TeBeest and Templeton, 1985). Most mycoherbicide projects currently in progress are following a similar pattern to the Collego study; that is, an initial feasibility or experimental phase (laboratory and greenhouse) which is funded by the public sector and terminates in small-scale trials to establish the efficacy of the fungus in the field. Private venture capital is then required to determine its market potential, formulate a product and then launch it in competition with chemical herbicides: this is the industrial or commercial phase. Sceptics, or simply those with little or no understanding of fungal epidemiology, may question the validity of the mycoherbicide approach. An obvious question is, if the pathogen is already present in the ecosystem then why does it fail to control the weed under normal circumstances? In the case of the Collego fungus, it can and does cause severe disease outbreaks (epiphytotics) in the field but typically only sporadically and late in the crop season when the weed has already had its major effect on yield and quality. The reasons for the slow and erratic disease build-up can be complex but essentially the Collego fungus has poor intercrop or interseasonal survival so the inoculum levels at the beginning of the growing season are low. Secondly, because it is a slime-spored fungus, aerial dissemination is inefficient and mainly restricted to short distance rainsplash, in sharp contrast to the dry-spored rust fungi, the dispersal ability of which has already been well demonstrated by the skeleton weed rust (*Puccinia chondrillina*). By applying massive amounts of inoculum (as formulated product) early in the season, almost total kill of the weed is achieved before there are any deleterious effects on the crop. Thus, the weak links (constraints) in the pathogen chain (life cycle) are overcome: namely, poor survival and dispersal ability. There is no need for further applications, thereby reducing costs as well as environmental pollution. Collego, therefore, is a good example of a successful mycoherbicide, both from an ecological and financial viewpoint. Does that signal the end for chemical herbicides? Unfortunately, nature is never that simplistic, the above example is for a single weed in a restricted cropping system which required alternative methods of control. Mycoherbicides are not the panacea; they will not replace chemical herbicides for many weed problems, either because there are no suitable pathogens or, ironically, because the pathogens are too specific or restrictive and control only a single weed species in a multi-weed situation. However, in certain weed – crop associations mycoherbicides may provide the answer where the use of chemical herbicides is environmentally or economically prohibitive. This will be discussed further within the context of grassy weed control.

Tropical grassy weeds

General projects

Of 18 weeds classified in Group 1 of the world's worst weeds (Holm *et al.*, 1977), 13 are grasses or sedges, seven of which have a tropical distribution. The natural enemy complex and the effect of that complex on the population dynamics of these grassy weeds is largely unknown and there is even a general lack of basic botanical information on the centres of origin and diversity of the majority of them. At present, most emphasis appears to be focused on control of Johnsongrass (*Sorghum halepense*) and *Cyperus* spp. in the USA. In contrast to other tropical grassy weeds, there is a relative abundance of literature on potential biological control agents, nearly all relating to mycoherbicide development. Chiang *et al.* (1986) evaluated the effect of numerous isolates belonging to four hyphomycete genera (*Bipolaris, Curvularia, Drechslera* and *Exserohilum*) both on leaf-pieces and whole plants of Johnsongrass. Complete kill of 3- to 4-week old plants was achieved by spraying conidial suspensions onto the leaves. Maize and sorghum both showed adverse symptoms but recovered from the initial injury. More recently, Mitchell (1988) screened *Colletotrichum graminicola* (Ces.) Wils. and *Gloeocercospora sorghi* Bain. & Edg. under controlled environment conditions against *Sorghum halepense* and other *Sorghum* spp. Only *C. graminicola* proved to be sufficiently specific to warrant further investigation as a potential mycoherbicide. Winder (1988) has field-tested *Bipolaris sorghicola* against the same weed with varying results. For the initial trial, a crude mycoherbicide preparation was applied during drought conditions but a 73% kill of seedlings was still obtained. However, subsequent tests proved to be disappointing and it was concluded that the effect of the pathogen is governed by weed density and soil type. *Dactylaria higginsii* (Luttrell) M.B. Ellis has been known for some time to cause epiphytotics in stands of *Cyperus rotundus* in the USA (Luttrell, 1954). However, it is only recently that the biological control potential of this fungal pathogen has been investigated. *D. higginsii* has since been recorded from Israel during searches for indigenous fungal pathogens of purple nutsedge (Bonne-Nitzani *et al.*, 1988). Severe disease outbreaks caused by this fungus have also been observed in Brazil in populations of *C. rotundus* (Barreto, pers.comm.). The possible exploitation of this fungus and other pathogens of *C. rotundus* reported by Evans (1987a) from literature and herbarium searches will be discussed later.

A unique approach to control of yellow nutsedge (*Cyperus esculentus*) has been developed over a number of years in the USA (Phatak *et al.*, 1983; Callaway *et al.*, 1987; Phatak, 1988). The indigenous rust fungus, *Puccinia canaliculata* (Schw.) Lagerh., is employed or manipulated in an integrated weed management system. In order to induce early and damaging rust epiphytotics on the weed and to avoid the inhibitory effects of fungicides applied to control crop pathogens, the rust is maintained in the greenhouse on its host during the winter months. The infected plants are placed in the field during May just as the nutsedge is appearing in the crop. The fungus dehydrates roots, leaves and tubers, as well as inhibiting flowering, making the weed non-competitive. In combination with certain herbicides, it also significantly reduces nutsedge populations the following year (Callaway *et al.*, 1987). The same rust species is also considered as a potential biological control agent for *C. rotundus*, however, the strains of *P. canaliculata* currently being studied for yellow nutsedge control are specific to *C. esculentus* and do not attack purple nutsedge (Callaway *et al.*, 1985). These authors consider that it should be possible to develop, by sexual recombination, a more virulent strain, which can be used against both *Cyperus* weeds.

Studies have also been initiated recently to assess the potential of fungal pathogens to control two of the major tropical grassy weeds: *Imperata cylindrica* and *Echinochloa crus-galli.* Preliminary results have been reported and appear promising (Scheepens, 1987; Caunter and Wong, 1988).

The itch-grass project

This project is funded by the Overseas Development Administration, and is being undertaken jointly by IIBC and Long Ashton Research Station (LARS), to explore the potential of mycoherbicides in controlling the tropical grassy weed, *Rottboellia cochinchinensis. R. cochinchinensis* (itch-grass) is a serious weed of many tropical crops including maize, sorghum, sugarcane and soyabeans. In smallholdings, farmers depend on manual weeding but if weed growth is advanced, workers may refuse to go into the fields as the fibreglass-like needles on the leaf sheath can penetrate hands and clothing. There are chemical herbicide treatments which can control itch-grass. However, most of these treatments are not selective to the crops usually associated with the weed and often the chemical does not remain in the soil long enough to suppress the succeeding flushes of germination. The weed is causing concern in the New World because of its relatively recent incursion into this region from its Old World origins.

The centre of evolution of the weed is unknown but an African or Asian origin is suspected. A range of fungal taxa belonging to the dematiaceous hyphomycete genus *Curvularia* was isolated from material received from various tropical countries (see Table 3.1), where disease outbreaks have been recorded. Screening of these isolates is being undertaken in the UK using the IIBC quarantine greenhouse fitted with negative pressure

Table 3.1. Effect of selected species and isolates of pathogens on *Rottboellia* and non-target Gramineae
n.b: All inoculations are by spraying unless otherwise stated.

		Pearl millet		*Sugarcane*			*Rice*			*Maize*					
Isolate	*Rottboellia Zimbabwe*	*HB-3*	*7042*	*Ex W.1*	*Q107*	*Cadmus*	*Blue Bonnet*	*Deep water*	*IR 36*	*Unknown*	*F_1 Hybrid Jubilee*	*Chag Trinidad*	*Malaysia*	*LG11*	*H632*
Curvularia cymbopogonis W1517 ex Bolivia	++++	+++	+++	++	+++	–	+	0	–	++	–	–	–	–	
C. cymbopogonis 1888 ex Bolivia	++	–	++	++++	–	–	–	–	–	–	–	–	–	–	
C. cymbopogonis 1862 ex Bolivia	++++	–	–	++++	–	–	–	–	–	–	–	–	–	–	
C. bicolor 1929 * ex Zimbabwe	++++	+++	+++	++	+++	–	–	–	–	++	–	–	–	–	
C. heterostrophus W945 ex Papua New Guinea	+++++	–	–	+++	–	++++	–	–	–	–	–	–	–	–	
Curvularia sp.1 W937 ex Trinidad[†]	+++++	0	0	0	0	–	0	0	–	–	0	–	–	–	

Table 3.1. *Continued*

		Pearl millet		*Sugarcane*			*Rice*			*Maize*					
Isolate	*Rottboellia Zimbabwe*	*HB-3*	*7042*	*Ex W.1*	*Q107*	*Cadmus*	*Blue Bonnet*	*Deep water*	*IR 36*	*Unknown*	*F_1 Hybrid Jubilee*	*Chag Trinidad*	*Malaysia*	*LG11*	*H632*
Curvularia sp.2 W940 ex Trinidad	++++	+++	+++	+++	+++	–	0	++	–	–	–	–	–	–	
Phoma sp. W938 ex Trinidad	0	0	0	0	0	–	–	–	–	0	–	–	–	–	
Curvularia sp.3 W941 ex Trinidad**	+++++	+++	++++	–	0	0	–	–	–	–	++	–	–	–	
Curvularia 8705 ex Papua New Guinea	0	0	0	+	+	–	–	–	0	–	–	+	+	+	–
Curvularia sp. isolate (1) ex Madagascar	++	++	–	++	++	–	–	–	0	–	–	+	+	+	–
Curvularia sp. isolate (2) ex Madagascar	+++	+++	+++	+++	++	–	–	–	++	–	–	+++	+++	+++	–
Curvularia sp. ex Zanzibar	++++	+++	+++	+++	+++	–	–	–	0	–	–	++++	++++	+++	–
Pyricularia sp. isolate (1) ex Kenya	++	–	–	–	–	–	–	–	–	–	–	–	–	–	–

Table 3.1. *Continued*

Pyricularia sp. (2) ex Kenya[††]	+	0	0	0	0	–	–	–	0	–	–	+	+	0	–
Pyricularia sp. ex Kenya brushed	++	–	–	–	–	–	–	–	–	–	–	–	–	–	–
Cercospora sp. ex Madagascar § brushed	0	–	–	–	–	–	–	–	–	–	–	–	–	–	–

Key to Table 1:

- – not tested
- 0 no infection (no visible symptoms)
- \+ minor symptoms, small number of plants affected
- ++ few lesions, variable or erratic infection rate
- +++ consistent infection, numerous small lesions (leaf spotting) or fewer larger lesions
- ++++ severe infection, coalescing lesions, leaf tip necrosis and leaf death
- +++++ widespread damping-off, host death

Footnotes:

* Many tests were carried out with this isolate and gave highly inconsistent results; the most severe symptoms, including damping-off, were seen when very young *Rottboellia* plants were brush-inoculated.

† Subsequent tests have given negative results.

** These symptoms were distinct from other isolates; severe red and yellow streaking was seen and infection appeared to be systemic since new leaves were infected. However, subsequent tests have failed to produce visible symptoms on any of the test plants. Similar symptoms to those first observed have been found on non-inoculated plants suggested that some other factor was involved.

†† The spore concentration of the suspension used for this test was very low which may have led to an anomalous result. The test will be repeated with a more concentrated spore suspension.

§ A method for producing a sporulating culture of the *Cercospora* sp. is being investigated, isolates have reverted to a mycelial, non-sporulating phase.

facilities to contain the pathogens. Laboratory experiments are carried out initially to optimize sporulation of the candidate pathogens and concentrated spore suspensions are applied to seedlings and young plants of *R. cochinchinensis* and the graminaceous crop plants most at risk (maize, sorghum, millet, rice and sugarcane). Plants are maintained overnight in a specially designed dew chamber to optimize infection conditions. Treated plants are observed for a period of one month and assessed weekly against controls. Isolates are made from any resultant lesions in order to corroborate pathogenicity (Koch's postulates). In many cases, symptoms of leaf chlorosis may be visible less than 24 h after inoculation indicating that phytotoxins are involved in pathogenesis. However, the results of these preliminary screening tests, using unselected pathogens, proved to be disappointing since the isolates either attacked non-target crop plants or failed to give consistently high infection on itch-grass. There is no doubt that certain taxonomic groups, particularly *Curvularia* spp., produce secondary metabolites with fast-acting herbicidal properties. On reflection, these results were not entirely unexpected as the majority of pathogens received were collected from exotic (non-indigenous) localities that *Rottboellia* has only recently invaded. Obviously, these are plurivorous (polyphagous) pathogens adapting quickly to new hosts from other graminaceous weeds or crop plants. Itch-grass, and tropical grassy weeds in general, could, therefore, constitute an important reservoir for pathogens of graminaceous crops, adding to their pest status. As stressed earlier, however, it is first necessary to determine the natural enemies present in the target area before searches are made in indigenous habitats where co-evolved pathogens with acceptably high specificity levels are more likely to be encountered.

Surveys were subsequently undertaken in Kenya, where itch-grass (shamba grass) is well established, having many local dialect names, and is seemingly a plant of long-standing in the region. Species of *Cercospora, Pyricularia, Phomopsis* and *Coniothyrium* were collected and isolated from diseased itch-grass plants throughout Kenya. An undescribed taxon of *Phomopsis* proved to be sufficiently damaging and specific in the preliminary screening experiments to warrant further investigation. However, when additional varieties of crop plants were included in more exhaustive tests, certain cultivars of maize proved to be susceptible and this pathogen was subsequently rejected. No *Colletotrichum* infections were observed in East Africa but herbarium searches at the International Mycological Institute (an Institute of CAB International) showed that *Colletotrichum* had been recorded on this weed in India. Further surveys were then concentrated in Asia (Nepal, India, Sri Lanka and Thailand). Numerous isolates of *Colletotrichum* were collected from these countries and, after prolonged greenhouse screening, four have been selected for further evaluation in the field. Trials will be carried out in the countries of origin of the strains (Sri Lanka and Thailand). The role of the LARS group is to develop appropriate production technology in order to increase the virulence and viability of the selected strains and provide a formulated product suitable for spraying in the field.

In addition to this mycoherbicide approach, a second option for biological control has arisen. A head smut, *Sphacelotheca ophiuri* (P. Henn.) Ling was found to be present in Kenya, Sri Lanka and Thailand, often causing locally damaging epiphytotics. In a weed where seeds are the only means of propagation and perennation, an inflorescence disease would appear to be a highly promising biological control agent. The smut has not been reported from the New World and hence a classical introduction could be considered. Investigations are currently underway to elucidate the life cycle and host specificity of *S. ophiuri.*

Future prospects

The major tropical grassy weeds are discussed individually in order of worldwide importance (Holm *et al.*, 1977), using mainly information collated by Evans (1987a).

Cyperus rotundus (*Cyperus* spp.)

As previously detailed, there appear to be a number of promising candidate fungi suitable for mycoherbicide investigation. Moreover, the concept of manipulating an obligate pathogen, such as the rust *Puccinia canaliculata,* is an interesting and novel approach to the biological control of weeds. The work of Phatak *et al.* (1983) shows clearly that fungal epiphytotics can be successfully induced in the field early in the season, thus overcoming the natural constraints to disease development, and significantly reducing both vegetative and reproductive capacity of the weed. Whilst these results are encouraging, the maintenance of rust cultures during the intercrop period and the subsequent release of inoculum into the weed population requires relatively precise techniques which may be difficult to co-ordinate in subsistence agriculture. Nevertheless, in well-organized, agricultural situations, it could be a viable control method not just for sedges but for grassy weeds in general. Further rusts have been recorded on *C. rotundus,* such as *Puccinia conclusa* Thum. and *P. philippinensis* H. & P. Sydow, and are apparently confined to the Old World tropics. This, combined with their ability to provoke severe leaf necrosis, makes them ideal candidates for classical biological control, and their introduction into the New World could be considered. *Phytophthora cyperi* (Ideta) Ido also shows promise as a classical biological control agent in view of its purported highly restricted geographic range (Sudan) and the severe host damage observed on herbarium specimens (Evans, 1987a). It should be noted here that the concept of using a 'dangerous' crop pathogen like *Phytophthora*, is not new and a commercial mycoherbicide (Devine) based on *P. palmivora*, has had spectacular success for control of milkweed vine (*Morrenia odorata*) in the USA (Templeton *et al.*, 1979; TeBeest and Templeton, 1985; Charudattan, 1987).

The dematiaceous hyphomycetes *Dactylaria higginsii*, *Duosporium cyperi* Thind. & Rawla and *Cercospora caricis* Oud. all appear to be damaging pathogens and hence warrant exploratory screening to evaluate host specificity and *in vitro* behaviour for possible mycoherbicide development. Inflorescence diseases of *Cyperus* spp., caused by *Cintractia limitata* Clinton (Ustilaginales) and *Balansia cyperi* Edgerton (Clavicipitales), have also been reported. However, their use in biological control is limited since most propagation in sedges is by vegetative means. Hence their overall effect on weed population dynamics is probably minimal.

Echinochloa crus-galli

The only relevant work on this weed has been carried out in Europe. Scheepens (1987) employed the shoot pathogen, *Cochliobolus lunatus* Nelson & Haasis, in combination with a sub-lethal dose of the chemical herbicide, atrazine, to kill young plants of this pantropical grassy weed. The fungus on its own can cause seedling mortality but older stages recover from initial infection. This synergism between pathogen and chemical herbicide holds considerable promise for future commercial exploitation. Similar seedling pathogens have been collected by the author on several tropical grassy weeds but once plants have reached a certain stage they are able to grow away from the infection. These would seem to be suitable cases for investigation with pathogen – chemical herbicide combinations.

Sorghum halepense

Work in the USA on the biological control of this weed has already been highlighted but it is considered that more surveys should be conducted before the centre of origin and the full spectrum of pathogens attacking this host can be determined. Similarly, there is a lack of herbarium and field data concerning the pathogens of *Eleusine indica*, *Paspalum conjugatum* and *Echinochloa* spp. in general, and clearly much more basic information is required before any assessment of biological control prospects can be made. However a recent study of two leaf-spotting fungi, *Diplodia setariae* Saw. and *Pyricularia grisea* (Cke.) Sacc., on goosegrass (*E. indica*), has shown that there is potential for the use of fungal pathogens as biological control agents of this weed (Figliola *et al.*, 1988)

Imperata cylindrica

Caunter and Wong (1988) recently undertook surveys for foliar pathogens of *I. cylindrica* in Malaysia and generally found low disease incidence. However, several candidate fungi were isolated including a *Colletotrichum* sp. This has proved to be specific to *I. cylindrica* in preliminary host range tests but the need for more extensive screening was stressed before the full potential of the pathogen can be evaluated, particularly any effects on rhizome viability and the regenerative ability of this noxious perennial weed.

Evans (1987a) concluded that it was difficult to target other candidate agents because of conflicting or insufficient information. For example, confusion over the taxonomy of the rusts involved (*Puccinia fragosoana* F. Beltran, *P. imperatae* Poirault and *P. rufipes* Dietel amongst others) makes it impossible to determine the actual distribution of a particular species. Several of the rusts are widespread in Africa and absent or uncommon in southeast Asia and the potential benefits of introducing these rusts into the region should be explored. Whilst on this theme, it should be pointed out that with many classical biological control agents there can be conflicts of interest. In the case of *I. cylindrica*, this may be a useful plant in certain countries or communities, being utilized for grazing or thatch. Thus, the introduction of an exotic, 'uncontrollable' pathogen into the ecosystem needs to be carefully assessed and the pros and cons balanced.

The smut, *Sphacelotheca schweinfurthiana* (Thum.) Sacc., is well-represented in the Old World, mainly Africa, and could also be considered for introduction into those areas of southeast Asia where *I. cylindrica* is a particularly important pest problem. Significantly, perhaps, these obligate pathogens (rusts, smuts) are common on *I. cylindrica* in the Mediterranean region, where the plant is not a serious weed (Holm *et al.*, 1977). The general scarcity of representatives of the Ascomycotina and Deuteromycotina associated with this host (Evans, 1987a), suggests that field surveys have been wanting and that these should be directed towards East Africa, the purported centre of origin of *I. cylindrica.* Recent observations in Papua New Guinea (Prior, pers.comm.) indicate that the mycoflora on this weed is considerably richer than the herbarium records suggest and that the true origin of the plant has yet to be determined. Another approach to the biological control of *I. cylindrica* is highlighted by current investigations in Brazil (Barreto, pers.comm.), in which fungal isolates from *I. brasiliensis* could be screened against *I. cylindrica* for possible introduction into the Old World as 'new encounters', either as classical or inundative biological control agents.

Conclusions and recommendations

The following statement encapsulates what the conclusions and recommendations of this paper should be:

> ... the current status of biological weed control with plant pathogens may be aptly characterized as a fluid, progressive one with increased interest among plant pathologists, weed societies, public agencies, fermentation scientists and industries. The needs for additional inputs into weed control technology are apparent to all. A glimpse of the potential of plant pathogens in weed control can be envisioned from the empirical success of a few.
>
> (Templeton, 1982)

This chapter has tried to show that, whilst the potential of fungal pathogens as biological control agents of tropical grassy weeds has not been realized, the prospects appear promising. Clearly, there is an urgent requirement for more basic research to explore this potential which can only be achieved by public sector funding, be it from state or international agencies. Once suitable agents have been identified, further funding for classical introductions would be minimal whilst that for mycoherbicide development would need to come from industry. Accordingly, agrochemical companies should be invited to participate in this phase of the investigation. Indeed, the continuing use of chemical herbicides, at greatly reduced concentrations in conjunction with pathogens, is reflected in the work of Scheepens (1987) and this is an exciting prospect suggesting a symbiotic relationship between traditional natural enemies; chemical companies and biological control organizations. A further link may be through the exploitation of the by-products of weed pathogens, using the secondary metabolites as 'environmentally friendly' chemical herbicides. There is evidence to support such a scenario. In the itch-grass project it was found that *Curvularia* isolates regularly produce lethal phytotoxins similar to those reported by Sugawara *et al.* (1985), who discovered that selective phytotoxins are formed in culture by strains of *Bipolaris cynodontis* (Marig.) Shoemaker, a pathogen of Bermuda grass in the USA.

The steps that ought to be taken in order to assess scientifically the potential of natural enemies to control tropical grassy weeds are as follows:

1. Develop a co-ordinated, international effort to enhance our understanding, and hence the exploitation, of biological control. This should involve arthropod natural enemies as well as pathogens since these still require investigation.
2. Basic research should be supported by funding from international agencies to investigate the potential of new endemic and exotic agents.
3. An international committee will be necessary to co-ordinate this

research, to actively seek funding and to develop guidelines for the safe use of biological control agents.

4. Because of the high specificity of many biological control agents, industry will be reluctant to invest in many of the projects, due to the restricted markets. Hence public sector funding will be required to promote small-scale or cottage industry production.

This chapter finishes with Simmonds' elegant and balanced plea which is still relevant to the current situation, particularly in relation to the biological control of tropical grassy weeds:

I would, therefore, suggest that far greater attention be paid to biological control in the tropics.... Certainly with present day ideas on pollution it is worthwhile avoiding unnecessary contamination in areas now comparatively free, and the use of biological control of pests is certainly helpful in this. This is by no means advocating abandoning the use of chemical pesticides since modern agricultural production depends very largely on them. I would, however, advocate consideration of their lessened but more effective use, and avoidance of misuse, in conjunction with adoption of biological control programmes as far as these are feasible – for the very obvious reason that in general they offer very high economic returns in lessening of crop losses from the funds invested.

(Simmonds, 1972)

References

Anon. (1989) *Agrow, Facts and Figures.* PJB Publications Ltd, Richmond, Surrey.

Bonne-Nitzani, E., Kenneth, R., Kleifield, Y. and Reuveni, R. (1988) Fungal diseases of noxious weed *Cyperus rotundus* (purple nutsedge) in Israel. *Proceedings of the Ninth Congress of the Israel Phytopathological Society,* Tel Aviv, February 1988.

Callaway, M.B., Phatak, S.C. and Wells, H.D. (1985) Studies on alternate hosts of the rust *Puccinia canaliculata,* a potential biological control agent for nutsedges. *Plant Disease* 69, 924–6.

Callaway, M.B., Phatak, S.C. and Wells, H.D. (1987) Interactions of *Puccinia canaliculata* (Schw.) Lagerh. with herbicides on tuber production and growth of *Cyperus esculentus* L. *Tropical Pest Management* 33, 22–6.

Caunter, I.G. and Wong, S.H. (1988) Foliar diseases of Lalang in Penang: biocontrol potential. *MAPPS Newsletter* 12, 6.

Charudattan, R. (1987) Devine leads the way. *Research 87, Biocontrol,* Institute of Food & Agricultural Science, University of Florida, pp38–9.

Chiang, Mou-Yen, Chilton, W.S. and Van Dyke, C.G. (1986) Laboratory and greenhouse screening of fungal pathogens with potential for control of Johnsongrass. *Proceedings of the Southern Weed Science Society, 39th Annual Meeting.*

Southern Weed Science Society, Champaign, Illinois, p386.

Cock, M.J.W. (1986) Biological control of weeds in tropical crops. In: Husseirn, M.Y. and Ibrahim, A.G. (eds) *Biological Control in the Tropics.* Penerbit Universiti Pertaman, Malaysia, pp375–85.

Cullen, J.M., Kable, P.F. and Catt, M. (1973) Epidemic spread of a rust imported for biological control. *Nature* 244, 462–4.

Cullen, J.M. and Hasan, S. (1988) Pathogens for the control of weeds. *Philosophical Transactions of the Royal Society of London* 318, 213–24.

East, R., Prestidge, R.A. and Robertson, L.N. (1986) Recent advances in pasture pest management in the northern North Island. *New Zealand Journal of Ecology* 9, 101–9.

Evans, H.C. (1987a) Fungal pathogens of some subtropical and tropical weeds and the possibilities for biological control. *Biocontrol News and Information* 8, 7–30.

Evans, H.C. (1987b) Life-cycle of *Puccinia abrupta* var. *partheniicola,* a potential biological control agent of *Parthenium hysterophorus. Transactions of the British Mycological Society* 88, 105–11.

Figliola, S.S., Camper, N.D. and Ridings, W.H. (1988) Potential biological control agents for goosegrass (*Eleusine indica*). *Weed Science* 36, 830–5.

Freeman, T.E. (1981) Use of conidial fungi in biological control. In: *Biology of Conidial Fungi Vol. 2,* Cole, G.T. and Kendrick, B. (eds) Academic Press, New York, pp143–65.

Girling, D.J. (1972) Investigation of the natural enemies of *Imperata cylindrica.* IIBC Unpublished Report.

Harris, P. (1988) Environmental impact of weed-control insects. *Bioscience* 38, 542–8.

Hasan, S. and Wapshere, A.J. (1973) The biology of *Puccinia chondrillina,* a potential biological control agent of skeleton weed. *Annals of Applied Biology* 74, 325–32.

Holm, L.G., Plucknett, D.L., Pancho, J.V. and Herberger, J.P. (1977) *The World's Worst Weeds. Distribution and Biology.* The University Press of Hawaii, Honolulu, 609pp

Julien, M.H. (1982) *Biological Control of Weeds: A World Catalogue of Agents and their Target Weeds.* CAB International, Wallingford, Oxfordshire.

Julien, M.H. (1987) *Biological Control of Weeds: A World Catalogue of Agents and their Target Weeds.* 2nd edn. CAB International, Wallingford, Oxfordshire, 144pp.

Lisansky, S.G. (1989) Wheat herbicides: the challenge of emerging resistance. *Pesticide Outlook* 1, 9–13.

Luttrell, E.S. (1954) An undescribed species of *Piricularia* on sedges. *Mycologia* 46, 810–14.

Mitchell, J.K. (1988) Evaluation of *Colletotrichum graminicola* (Les.) Wils. and *Gloeocercospora sorghi* D. Bann. & Edg. as biological herbicides for controlling johnsongrass (*Sorghum halepense* (L.) Pers.). *Proceedings of the VII International Symposium on Biological Control of Weeds,* Rome, March 1988 (Abstract).

Mortensen, K. (1986) Biological control of weeds with plant pathogens. *Canadian Journal of Plant Pathology* 8, 229–31.

Nye, I.W.B. (1958) The external morphology of some of the dipterous larvae living in the Gramineae of Britain. *Transactions of the Royal Entomological Society of London* 110, 411–87.

Phatak, S.C. (1988) Rust (*Puccinia canaliculata*) and nutsedges (*Cyperus* spp.). *Proceedings of the VII International Symposium on Biological Control of Weeds*, Rome, March 1988 (Abstract).

Phatak, S.C., Sumner, D.R., Wells, H.D., Bell, D.K. and Glaze, N.C. (1983) Biological control of yellow nutsedge with the indigenous rust fungus *Puccinia canaliculata. Science* 219, 1446–7.

Savile, D.B.O. (1954) The fungi as aids in the taxonomy of the flowering plants. *Science* 120, 583–5.

Scheepens, P.C. (1987) Joint action of *Cochliobolus lunatus* and atrazine on *Echinochloa crus-galli* (L.) Beauv. *Weed Research* 27, 43–7.

Schroeder, D. (1980) The biological control of thistles. *Biocontrol News and Information* 1, 9–26.

Schroeder, D. (1983) Biological control of weeds. In: Fletcher, W.W. (ed.) *Recent Advances in Weed Research.* CAB International, Wallingford, Oxfordshire, pp41–78.

Schroeder, D. and Goeden, R.D. (1986) The search for arthropod natural enemies of introduced weeds for biological control in theory and practice. *Biocontrol News and Information* 7, 147–55.

Simmonds, F.J. (1972) Biological control in the tropics. CIBC Technical Bulletin 15, 159–69.

Strong, D.R., Lawton, H. and Southwood, R. (1984) *Insects on Plants.* Blackwell Scientific Publications, Oxford.

Sugawara, F., Strobel, G., Fisher, L.E., Van Duyne, G.D. and Clardy, J. (1985) Bipolaroxin, a selective phytotoxin produced by *Bipolaris cynodontis. Proceedings of the National Academy of Sciences, USA* 82, 8291–4.

TeBeest, D.O. and Templeton, G.E. (1985) Mycoherbicides: progress in biological control of weeds. *Plant Disease* 69, 6–10.

Templeton, G.E. (1982) Status of weed control with plant pathogens. In: Charudattan, R. and Walker, H.L. (eds) *Biological Control of Weeds with Plant Pathogens.* J. Wiley & Sons, New York, pp. 29–44.

Templeton, G.E. and Greaves, M.P. (1984) Biological control of weeds with fungal pathogens. *Tropical Pest Management* 30, 333–8.

Templeton, G.E., TeBeest, D.O. and Smith, R.J. (1979) Biological control with mycoherbicides. *Annual Review of Phytopathology* 17, 301–10.

Wapshere, A.J. (1974) A strategy for evaluating the safety of organisms for biological weed control. *Annals of Applied Biology* 77, 201–11.

Wapshere, A.J. (1989) A testing sequence for reducing rejection of potential biological control agents for weeds. *Annals of Applied Biology* 114, 515–26.

Winder, R.S. (1988) Field testing of *Bipolaris sorghicola* as a mycoherbicide for johnsongrass. *Proceedings of the VII International Symposium on Biological Control of Weeds*, Rome, March 1988 (Abstract).

Worsham, A.D. (1982) Discussion of topics. In: Charudattan, R. and Walker, H.L. (eds) *Biological Control of Weeds with Plant Pathogens.* J. Wiley and Sons; New York, pp219–36.

Chapter 4
Chemical Control of Grassy Weeds

S.C. Collins

Monsanto Central Africa Inc.
P.O. Box 47686, Nairobi, Kenya

Introduction

The earliest grass weed control was obtained with the triazines to a limited extent. As far back as the end of the 1950s the urea derivatives were developed and in the early 1960s the bipyridilium herbicides with their desiccation effect were successful in control of early annual grasses prior to planting. Next came the dithiocarbamates followed by the first specific grass herbicides. Tri-allate was developed in 1964 for control of *Avena* in small grain cereals (see Tables 4.1 and 4.2).

Table 4.1. History of graminicide development. Decades when products were announced and subsequently developed.

Years	*Products*
1940s	2, 4-D, DNOC, MCPA, propham, TCA
1950s	Atrazine, dalapon, di-allate, diquat, diuron, EPTC, paraquat, simazine
1960s	Alachlor, asulam, bentazone, chlortoluron, cyanazine, dichlobenil, ethofumesate, ioxynil, metribuzine, tri-allate, trifluralin
1970s	Difenzoquat, glyphosate, pendimethalin
1980s	Bensulfuron-methyl, chlorsulfuron, esprocarb, glufosinate-ammonium, imazapyr, imazaquin, metazachlor, triclopyr

Table 4.2. Post-emergence graminicides: years of introduction, chemical groups and manufacturing companies.

Year	*Chemical group*	*Common name*	*Manufacturer*
1970	Halogenated alkanoic acid	Dalapon-sodium	Dow
1971	Organophosphorus	Glyphosate	Monsanto
1974	Quaternary ammonium	Difenzoquat	American Cyanamid
1975	Substituted urea	Isoproturon	Hoechst, Ciba-Geigy, Rhone-Poulenc
1976	Phenoxy acid	Diclofop-methyl	Hoechst
1977	Phenoxy acid	Fluazifop-butyl	ICI
1980	Sulfonyl urea	Chlorsulfuron	Dupont, Ciba-Geigy, Ishihara, Nissan
1981	Organophosphorus	Glufosinate-ammonium	Hoechst
1981	Pyridine	Triclopyr	Dow
1983	Imidazolinone	Imazapyr	American Cyanamid
	Imidazolinone	Imazaquin	American Cyanamid
1988	Phenoxy acid	Fenoxaprop-ethyl	Hoechst

The acetanilides were the first major group of selective grass herbicides. Whereas the triazines were predominantly for dicot control, the acetanilides, specifically alachlor, were more monocot-specific and were developed for use in maize, soyabean, cotton, brassicas and sugarcane. Alachlor was followed by metolachlor and others.

In the 1960s the dinitroanilines were introduced, including trifluralin which is volatile and needs incorporation, and later pendimethalin for a broad range of annual grasses in cereals and other crops. Towards the end of the 1970s a large group of selective graminicides was developed, mostly for post-emergence use. These were more selective and were active at lower rates of active ingredient per ha than previous products. The history of graminicides and specific graminicides is summarized in Tables 4.1 and 4.2.

There is a wide range of graminicides now on the market which when used singly, or in combination, have sufficient efficacy, weed spectrum, persistence, residuality, degradation and selectivity for most grass weed–crop situations. There is room for improvement for economic sedge control, *Bromus* control in small grain cereals and some aggressive broadleaved species in fallow situations.

Graminicides provide the most efficient form of weed control but they all have their particular scope and limitations which must be taken into account for any particular crop–weed situation. In general, graminicides have been well researched and provide a reliable, safe and convenient form of weed control when used according to label directions.

Mode of action of some important graminicides

Graminicides are usually either taken up by foliage (post-emergence) or by roots or other below-ground organs (pre-emergence). Post-emergence herbicides are dependent on spray retention on the leaf which is determined by the characteristics of the leaf itself, the spray volume and droplet size, surfactants and weather. Soil herbicides are dependent on the characteristics of the soil surface, soil moisture, soil texture and the relationship between the placement of the herbicide and its effect on the plant. An additional factor is the time of application of the herbicide in relation to the germination of the weeds and crop.

Triazines (atrazine, simazine, cyanazine)

The triazines are the most widely used herbicides in the tropics. They show a high degree of crop tolerance and soil activity, the latter being influenced by their solubility in water. Uptake is mainly through the plant roots while foliar uptake occurs with the more soluble triazines (ametryn, atrazine, cyanazine, prometryn and terbutryne). The triazines move in the transpiration stream mostly in the apoplast of treated plants and inhibit photosynthetic electron transport.

Thiocarbamate herbicides (EPTC, butylate)

These are volatile and require incorporation. Butylate is readily absorbed by plant roots and translocated to the above ground parts of the plant. It inhibits growth in grass meristems but is metabolized in roots and leaves of maize. Breakdown is microbial.

Acetanilides

These products (alachlor, metolachlor, propachlor and butachlor) are active at 2–4 kg/ha rates and are generally selective pre- and early post-emergence. They are absorbed by germinating shoots, translocated throughout the plant to vegetative parts and inhibit root and shoot development. Germinating grasses absorb the chemical through the coleoptile or shoot tip. Degradation is microbial with a soil persistence of 40–70 days.

Dinitroaniline herbicides

Many in this group are soil active with little or no foliar activity. They tend to volatilize when applied on the soil surface and so need incorporation. Oryzalin and pendimethalin are two members of this group that are less volatile and can be applied pre-emergence without soil incorporation. Some soil moisture is required by pendimethalin for uptake by germinating weeds and its low solubility ensures that it remains in the top layer of soil. These herbicides kill seedling weeds by inhibiting the development of lateral roots in susceptible plants.

Phenoxy herbicides

This is a large group that originated with the broadleaf herbicides 2, 4-D and MCPA. More specific action is demonstrated by diclofop-methyl, fenoxaprop-ethyl and fluazifop-butyl which are graminicides with no effect on broadleaves. Fenoxaprop-ethyl is taken up by the plant tissues and is transported to the growing points of leaf bases and roots where inhibition of fatty synthesis takes place.

Sulfonylureas

These herbicides are applied pre-emergence or post-emergence because they are taken up by the roots or foliage. The solubility of chlorsulfuron varies with pH, being low under acidic conditions and high in alkaline conditions. It is non-volatile and the principal degradation pathway is by hydrolysis to non-herbicidal compounds. There is rapid acropetal and basipetal translocation followed by rapid growh inhibition at both root and shoot tips. Cell division is stopped at the growing points and is visible after a few days. Plants either die or remain stunted.

Bipyridilium herbicides

Paraquat and diquat having desiccant activity are both foliar acting contact herbicides with no soil activity and so have to be applied post-emergence. Injury symptoms are very rapid: all green tissue exposed to the herbicide being injured. The action of the herbicides is dependent on light and they interfere with photosynthesis by diverting electrons from ferridoxin. They are very strongly adsorbed to soil and are therefore not taken up by roots.

Organophosphorus herbicides

Glufosinate-ammonium is a partially systemic contact herbicide. It inhibits the action of the enzyme glutamine synthetase leading to the release of ammonia which is phytotoxic. It is active at 1–2 kg a.i./ha. Glyphosate is, however, rapidly translocated to all parts of the plant both above and below ground. It acts on aromatic amino acid biosynthesis by mimicking an amino acid in the shikimic acid pathway. It is strongly adsorbed to soils, is not taken up by roots or seeds and is rapidly broken down microbially. Its systemic mode of action, low mammalian toxicity and uptake through passive diffusion confer exceptional suitability for low volume application along with its associated logistic benefits.

Imidazolinones

The mode of action is by inhibition of amino-acid synthesis and symptoms are slow to appear. The rate of degradation depends on temperature and soil type, the compounds taking longer to degrade in cooler climates.

Efficacy of some important graminicides

- Triazines control seedling grasses and broadleaves.
- Thiocarbamates are more effective on germinating grass weeds than on established grass, with some efficacy on *Cyperus rotundus.*
- Acetanilides generally need to be mixed with a broadleaf herbicide such as a triazine in maize or with a urea for legumes to ensure sufficiently wide spectrum weed control. Acetochlor is more active than alachlor but can be more phytotoxic to crops at higher rates on light soils. A safener overcomes this problem. Butachlor is selective in transplanted rice and is used to control *Echinochloa* spp. and *Cyperus difformis*.
- Phenoxy acids: diclofop-methyl controls annual grasses with virtually no effect on broadleaves or perennial grasses. It is used post-emergence on annual grasses in the seedling stage. Fenoxaprop-ethyl is effective on most annual grasses up to the first node growth stage with some effect on perennials. Fluazifop-butyl, chlorazifop-propynl and haloxyfop-methyl are phytotoxic to grasses while broadleaves are unaffected.
- Dinitroanilines are effective in the control of *Rottboellia cochinchinensis*. Pendimethalin has a very narrow weed spectrum but it can be broadened by tankmixing with atrazine for use in maize. The grass weed susceptibility covers many important annual grass weeds and is listed as follows:

  Setaria verticillata (L.) P. Beauv.
  Setaria pumila (Poir.) Roem and Schult.

Eleusine indica (L.) Gaertn.
Eleusine indica subsp. *africana* (Kennedy–O'Byrne) S.M. Phillips.
Poa spp.
Rottboellia cochinchinensis (Lour.) W.D. Clayton.
Sorghum halepense (L.) Pers.

Notable grass weeds not usually controlled by pendimethalin are as follows:

Eleusine multiflora A. Rich.
Bromus spp.
Lolium spp.
Avena spp.

• Sulfonylureas all have very high unit activity, 15–30 g a.i./ha of chlorsulfuron being sufficient in most situations.

• Bipyridilium herbicides: paraquat has a wide spectrum of weed control in that it affects all green tissue that it contacts. It has limited efficacy on perennial weeds but is more effective on weeds which are small and in an early establishing or vegetative phase of growth. Some annual grasses may only be temporarily suppressed because the low and enclosed growing points are not contacted by the spray. The effect is similar to that of grazing.

• Glufosinate-ammonium is partially systemic, giving more persistent control of grasses than paraquat, but not as effective as glyphosate.

• Glyphosate: the complete translocation of this herbicide confers remarkable efficacy on most weeds, whether they are annual broadleaves and grasses, perennial broadleaves and grasses or sedges. Symptom development is slow but irreversible.

• Imidazolinones: a high level of phytotoxicity permits low dose rates (0.15–0.75 kg a.i./ha). Imazamethabenz is specific for wild oat control. Imazaquin normally needs a specific graminicide in tank mixture. Imazamethabenz and Imazethapyr are sometimes formulated with pendamethalin to widen or strengthen grass control. Imazapyr controls sedges as well as many perennial grasses at rates higher than those required for broadleaves.

• Substituted ureas: isoproturon is active on annual grasses *Alopecurus*, *Avena* and *Poa* as well as broadleaves in barley, rye and wheat at 1–1.5 kg a.i./ha.

Crop selectivity

Herbicides appropriate for a range of crop and grassy weed situations are listed in Table 4.3.

- Dinitroanilines: pendimethalin can be used in many crops ranging from maize, sorghum, cowpea and soyabean to vegetable crops. Pendimethalin's selectivity in most crops is based on enzymatic degradation but in rice it is dependent on seed placement in the soil. It causes stand reduction in shallow planted dry rice but not in deep planted rice. Trifluralin can be used in a wide range of crops but its requirement for soil incorporation limits its use.

- Thiocarbamates: selectivity on maize is acceptable, particularly if a safener is added.

- Acetanilides: pre-emergence and early post-emergence selectivity is obtained with alachlor, metolachlor and acetochlor in maize, soyabean, beans, groundnuts, potatoes, and sugarcane. Alachlor can be used in cotton in some cases.

- Phenoxy acids: diclofop-methyl is selective in wheat, barley, beans, soyabeans, peas and sunflowers. Sometimes there is a yellowing effect on wheat when applied at an early stage but the plant quickly grows out of this and there is no effect on yield. Fluazifop-butyl, chlorazifop-propynyl, fenoxaprop-ethyl, sethoxydim and haloxyfop-methyl are all post-emergence graminicides for use on broadleaf crops. Fenoxaprop-ethyl can be used for annual grass control in wheat at 90–120 g a.i./ha.

- Sulfonylureas: This group has 4 very diverse selectivity properties as shown in Table 4.4. Soil residues of chlorsulfuron have sometimes damaged beans, maize and other crops.

- Bipyridilium herbicides: These being desiccants are not selective and need to be applied prior to planting any crop or as a directed (tree crops) or shielded application within the crop inter-row (e.g. maize).

- Glufosinate-ammonium is a post-emergence non-selective herbicide and, like paraquat and glyphosate, needs to be applied as a preplanting, directed or shielded spray to avoid desirable foliage.

- Glyphosate is non-selective and should be used preplant or as a directed or shielded spray in crops. Directed, unshielded applications are possible in mature coffee, tea and citrus which are relatively resistant to glyphosate. Glyphosate is also registered as a preharvest treatment in some areas to be used on small-grain cereals, cotton or soyabean which have reached physiological maturity. Glyphosate weed control programmes at very low rates have provided very economic and convenient weed control in plantation crops.

- Imidazolinones: Imazamethabenz is selective in wheat and barley. Imazaquin is selective in cowpea, beans, soyabean, tobacco and coffee and is phytotoxic to maize, sorghum and cotton. Imazethapyr is selective on

Table 4.3. Crop–grassy weed situations.

Situation	*Herbicide*	*Time of application*
Non-crop/fallow/pre-plant	Glyphosate	post
	Atrazine	pre
	Paraquat	post
	Glufosinate-ammonium	post
Avena spp. in cereals	Tri-allate	ppi
	Difenzoquat	post
	Benzoylprop-ethyl	post
Annual and perennial grasses in plantation crops	Glyphosate	post
	Simazine	pre
	Imazapyr	post
	Trichlopyr	post
	Paraquat	post
Annual and perennial grasses in orchards and vines	Aminotriazole	post
	Simazine	pre
	Glyphosate	post
	Paraquat	post
Annual grasses and sedges in rice	Propanil	pre/post
	Butachlor	pre
	Esprocarb-bensulforon-methyl	post
	Oxadiazon	pre
Annual grasses in cereals	Chlortoluron	post
	Isoproturon	post
	Imazaquin	post
Annual grasses in maize	Atrazine	pre
	Alachlor	pre
	Metolachlor	pre
Annual grasses in soyabean	Alachlor	pre
	Metolachlor	pre
	Propachlor	pre
	Oryzalin	pre
	Imazaquin	pre
	Quizalafop	post
	Fluazifop-butyl	post
	Barban	post
	Diclofop-methyl	post
	Pendimethalin	pre
Annual grasses in cotton	Pendimethalin	pre
	Trifluralin	ppi
	Norflurazon	pre

Situation	*Herbicide*	*Time of application*
	Fluazifop-butyl	post
	Quizalafop	post
Annual/perennial grasses in sugarbeet/sunflower/potato	Metamitron	pre
	Fluazifop-butyl	post
	Quizalafop	post
Annual/perennial grasses in sugarcane	Asulam	post
	Dalapon	post
	Metribuzin	pre
	TCA	pre
	Acetochlor	pre
Annual grasses in herbage grass	Ethofumusate	post
	Diclofop-methyl	post

legumes. Imazapyr is non-selective and is used in reforestation sites and non-crop land.

Weed flora and herbicide resistance

Any weed control method will tend to change the weed spectrum. Hand methods have tended to increase the prevalence of perennial weeds while slashing tends to favour the predominance of grasses. The use of chemicals has a similar effect; atrazine use led to grass weed problems such as *Setaria* spp. which then required the development of specific grass herbicides.

Continued use of single products with single target site activity over a period of many years can lead to the development of herbicide resistance. The majority of occurrences have been in North America but other parts of the world are also affected. Examples are as follows:

- *Setaria viridis* (L.) P. Beauv. Trifluralin resistance in Canada;
- *Lolium rigidum* Gaud. Diclofop-methyl resistance in Australia;
- *Digitaria* spp. and various *Amaranthus* spp. are resistant to atrazine in Europe;
- Dicot resistance to sulfonylurea herbicides in USA.

A distinction should be made between resistance and tolerance which can be drawn as follows:

Tolerance: A plant's capacity to withstand herbicide treatment at normally used doses without injury or lasting damage to its growth or development.

Table 4.4. Sulfonylurea crop spectrum.

Common name or code	*Crop and uses*
Chlorsulfuron	Cereals (pre/post) Flax (post)
Metsulfuron-methyl	Cereals (post) Rice (post) Rubber and oil palm
DPX-L5300	Cereals (post)
DPX-M6316	Cereals (post) Maize (post)
Chlorimuron-ethyl	Soyabean (post) Sugarcane (post)
Bensulfuron-methyl	Rice (post)
Sulfometuron-methyl	Non crop (pre/post) Forestry

Resistance: Ability of a plant to grow normally in spite of exposure to normally used doses unlike a tolerant plant. The resistant plant does not depend on preventing the entry of the herbicide to which it is resistant but lacks a metabolic site for the action of the herbicide.

Development of resistance in low input agriculture, such as that of many tropical countries, is unlikely. Atrazine resistance has been well established in Europe after some 25 seasons' continual use but resistance to sulfonylureas developed more rapidly. In the tropics, high temperature and humidity speed up the degradation of plant protection products.

GIFAP, the agricultural chemical industry association, has set up a Herbicide Resistance Action Committee (HRAC) to see if herbicide resistance can be prevented or slowed down. Initially there are three working groups responsible for (a) triazines, graminicides and sulfonylureas, (b) imidazolinones and (c) triazolpyrimidines.

New chemistry

A number of new sulfonylureas are under development, such as DPX-E9636 and SL-950 for maize and RH-0898 for post-emergence grass control in wheat, rice and broadleaf crops. UBI C-4874 (MON 31000) is a new phenoxypropionate compound for sugarbeet and oilseed rape and CGA 184'927+S(S is a new safener) for cereals. All these products are performing at less than 500 g a.i./ha. HOE 70542 is a triazole-based fenoxaprop safener for *Avena* in cereals which was announced at the 1989 Brighton conference.

Future directions

The cost of research and development of chemicals is escalating, associated with the requirement for more comprehensive toxicological packages. It is to be expected that more mixtures with existing products will be developed. A number of products will be restricted on toxicological grounds as a result of more stringent legislation and the pace of new product introduction will be very much reduced.

Issues such as groundwater pollution will gain momentum and products such as atrazine will be restricted in developed countries. Cumbersome legislation, freedom of information together with the environmental anti-plant protection lobby will bring additional restrictions or bans in developing countries where these same compounds may best meet the needs of the local situation. There is a risk that newly established registration authorities will make decisions without fully understanding the implications of the local situation.

New compounds having higher unit activity and superior handling properties will be more efficient and more environmentally safe. High unit activity means less product being applied, less waste and fewer residues. New compounds will have improved toxicological and environmental properties.

It is an undoubted fact that pesticides, including graminicides, will play a major part in crop protection for many years to come. They will have an essential role to fill in ensuring efficient and effective weed control whilst achieving a higher standard of living for growers, who would otherwise have to spend most of their time hand-weeding. It is estimated that the average African farmer is forced, quite often, to spend more than half his working time on weeding with hands or hoes. Weeds in tropical agriculture destroy between 25% and 100% of the harvest if not properly managed. Chemical weed control represents an efficient and highly effective manner of optimizing yield and quality, thereby freeing the farmer for other priority tasks.

Biotechnology offers more exciting possibilities with the development of herbicide-resistant crop varieties such as sugarbeet, cotton, soyabean and tobacco. Genetic engineering will offer new plant protection opportunities provided decisions are based on scientific rather than emotional grounds. Research trends will still be geared to the world's major food and cash crops, including maize, small grains, soyabean, cotton and various broadleaf crops.

Useful literature

Akobundu, I.O. (1987) *Weed Science in the Tropics: Principles and Practices.* Wiley, 522pp.

Hance, R.J. and Holly, K. (eds) (1990) *Weed Control Handbook: Principles.* Blackwell, 582pp.

Klingman, G.C., Ashton, F.M. and Noordoff, L.M. (1982) *Weed Science: Principles and Practices*. J. Wiley & Son.

Worthing, C.R. and Walker, S.B. (eds) (1987) *The Pesticide Manual, A World Compendium.* British Crop Protection Council, 1081pp.

Chapter 5
Safe Use of Pesticides on Grassy Weeds

E. Dorn

Hoechst, Postfach 800320,
Frankfurt-am-Main 80, Germany.

Introduction

At the very beginning of the *Resolution 10/85 of the 23rd FAO Conference of November 1985 International Code of Conduct on the Distribution and Use of Pesticides* it is stated:

The conference recognize(s) that increased food production is a high priority in many parts of the world and that this need cannot be met without the use of indispensable agricultural inputs such as pesticides ...

and further:

Acknowledging that pesticides can be hazardous to humans and the environment and that immediate action must be taken by all concerned, including governments, manufacturers, traders, and users, to eliminate as far as possible and within the scope of their responsibility, unreasonable risks, not only in the country of origin but also in the countries to which pesticides may be exported.

These words describe precisely the environment of the subject discussed below. We need more food and to achieve this goal we need to apply safely more pesticides that are intrinsically safe.

This chapter describes how safe pesticides can be selected from many pesticide candidates at the screening stage, and can be developed as a safe tool for the farmer to use, to solve his serious pest, disease and weed control problems during food production. A pesticide is defined as safe when the information, the measures, as well as the techniques are available to handle it safely. To express it negatively, when safety instructions are not followed even a safe pesticide may be misused, and may cause adverse effects to humans and the environment. In addition, pesticides can be packed, shipped, stored, distributed and, highly important, applied in a way which is acceptably safe for humans as applicators and as consumers of agricultural products, and equally safe for non-target organisms in the environment. This chapter outlines the strategy and the validated standard tests for investigation of the compound properties. These tests have been developed for selecting those candidates in many varied screening steps, which fulfil all the safety demands established by the scientific and regulatory community.

Scenario of pesticide production and use

When a team of experts develops new pesticides, it is important that those who are responsible for all safety aspects know the scenarios of pesticide handling from the production plant through to its use in the field, and finally the disposal of pesticide containers and residual non-used portions. They must be aware of all possible routes of pesticide translocation and contamination.

As pesticides are physiologically active substances *per se* it is not sufficient to prevent visible contamination, because this level of exposure may already cause toxic effects. Furthermore, it is not acceptable to allow contamination until the first intoxication, such as when skin irritation or headache is first observed. Therefore these experts must have 'trace analytical glasses' to recognize those minute amounts which reach off-target sites even when good manufacturing and agricultural practices have been applied.

Figure 5.1 shows schematically the processes of translocation. The safety expert needs to know the particular sites of potential entry into the human body, e.g. the skin, the lung for airborne contamination and the alimentary canal for food which may carry low levels of residues. He must follow the translocation of pesticide residues in the environment to non-target organisms via drift during spraying, run-off from the field with rain-water, post-application volatilization and transport in the air, as well as penetration into groundwater through the soil, which may finally end up in drinking water.

In addition to these routes of translocation, the safety expert needs to determine the composition of residues from the intact parent active ingredient and its degradation products in all those matrices which finally may come into contact with humans and non-target organisms. This means the chemical, photochemical and biochemical conversion products, which are present at any relevant period of time and level, must be determined by the analytical expert. Only then will he know the total exposure. The expert has to decide, whether his basic set of data, generated by investigating re-

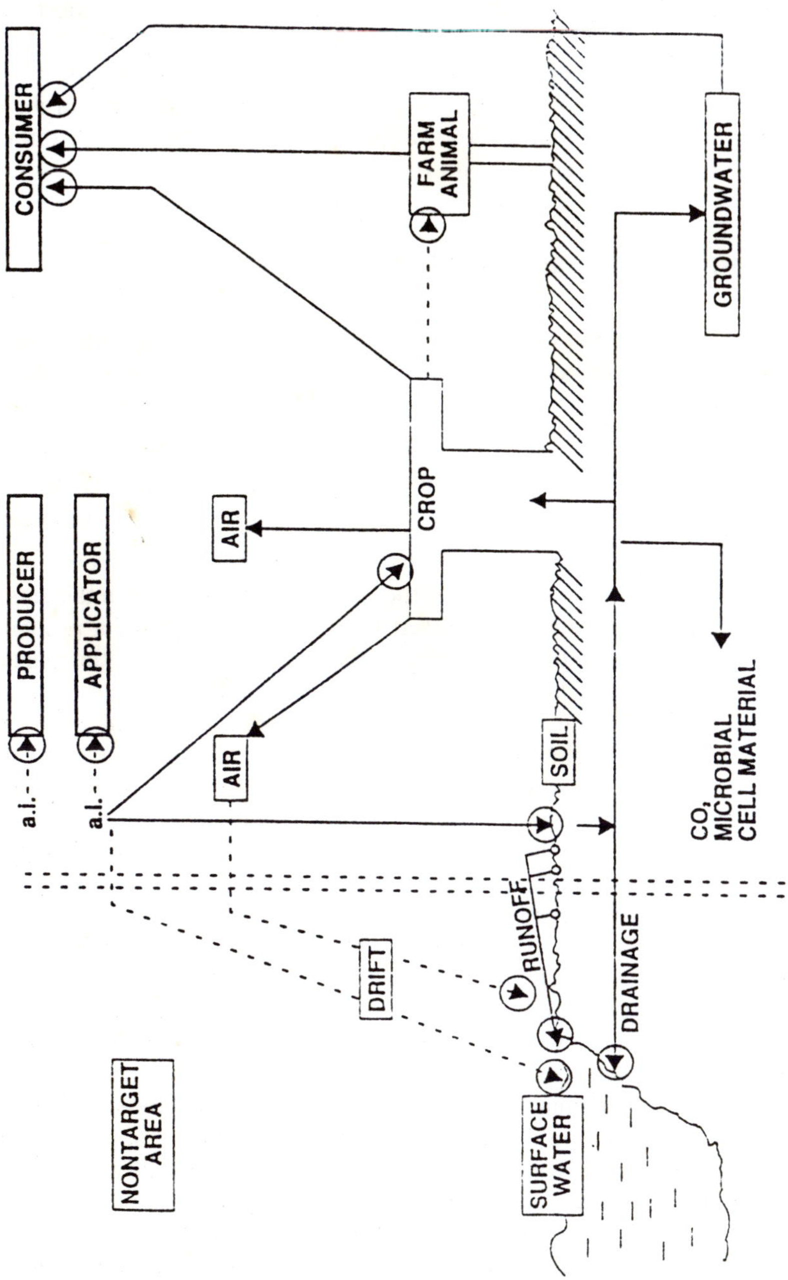

Fig. 5.1 Processes of translocation of pesticides.

presentative scenarios of pesticide handling, is sufficient to describe the total exposure under different conditions, or whether specific additional tests must be carried out under different climatic conditions.
With this essential information he can now ask two subsequent questions.

First, what degree of hazard – or reduction of hazard – does this exposure level represent, and if necessary how can one reduce the exposure to an acceptable level? Secondly, we must establish and maintain a margin of safety between non-avoidable exposure and unacceptable hazard. Otherwise the pesticide candidate can not be licensed.

Definition of safety and risk for man and environment

Within the context of the risk and safety of pesticide handling there are two expressions for one aspect which are inversely proportional to each other. When measures reduce risk, safety increases. The official definition of risk reads:

Risk = (Probability of the occurrence of a damaging event) × (Extent of damage)

This definition reflects very well the two aspects which are relevant in risk assessment. First, the position of the human or non-target organisms is mentioned where they may be exposed to a particular level of a pesticide residue. This describes the probability of damage. Secondly, the hazard itself, the nature of the toxic effects, and in particular at which exposure threshold level intoxication may be observed. Another quantitative expression for risk is the margin or distance between the exposure level, and the threshold level, above which damage occurs, and below which safety begins. Due to the variation and some uncertainties in the determination of these two borderlines, i.e. highest exposure possible and lowest toxicological threshold level, the margins of safety (MOS) are defined and established to ensure safety even in extreme situations of pesticide handling.

This approach is generally applied to thousands of risk factors to which we are exposed every day such as traffic, accidental situations at home and at work, during sports activities, administration of pharmaceutical drugs, unbalanced nutrition as well as a too high or too low food consumption. However, a complete risk assessment is always symmetrical. When assessing the risk of regular sports activity, one must also assess the risk on health of not participating in any sport. The same has to be applied to the risk of using pesticides. Not using pesticides would dramatically reduce the food supply in many countries with the consequence of an unbalanced diet, severe health effects, and possibly starvation. As a second consequence, the growth of fungi on food commodities can be envisaged, followed by the formation of mycotoxins up to concentrations which cause health defects.

Necessary information for the assessment of risk

As mentioned earlier, the experts responsible for the safety aspects of pesticide handling must be made well aware of all the possible scenarios and circumstances. These experts from agencies, the chemical industry and academia define the relevant segments of this overall scenario, or in other words break down the sequence of events and operations into separate steps, from production and packaging, transport, mixing and spraying, to harvesting and the processing of raw agricultural commodities. Then adequate experimental designs and protocols are developed. These protocols clearly define the prerequisites needed in order to properly establish the experimental design. The studies include determinations of the metabolic pattern of residues in the mature crop plant, and finally how to analyse all the parameters of interest.

These models fulfil the two requirements: (i) to be standardized in order to generate reproducible results; and (ii) to represent the real situation in the field. In the schematic diagram of Fig. 5.1, all arrows represent a translocation process from one compartment to another. At each arrow point three parameters are determined: (a) nature of residues (metabolites) in the matrix given; (b) the concentrations of these residues at all relevant time stages; and (c) the non-toxic threshold level (susceptibility) of the respective non-target organism which might be exposed to residues via a given route.

For each of these parameters a representative test system has been developed and most are subject to worldwide agreements. Humans cannot be tested directly for their susceptibility to pesticide exposure. Therefore, representative animals are used to determine the crucial parameters, such as the non-toxic threshold levels below which no adverse effects can be observed. The extrapolation of data on the behaviour of residues (metabolism and toxic action) in animals to humans might contain a certain degree of error. This, however, is limited as we know from a vast range of experience. Three major arguments may clarify this further. First, at least three different animal species are tested for their susceptibility and the most susceptible one is considered for risk calculations. With this assumption a degree of safety is implemented. However, if there are great differences in the toxicological threshold levels of different species, toxicologists elucidate the reasons for these effects and have to define which species is likely to have the most similar reactions to man. This one is then considered for the final risk evaluation. Second, because of the uncertainty of possibly higher susceptibility in man in comparison with the most sensitive test animal, far lower exposure to humans is permitted – generally by a factor of 100 lower – than the 'no observable' effect level in the most susceptible animal. Third, the vast experience with pharmaceutical drugs, the behaviour of which is tested in animals and also in humans, confirmed the assumption that the standard test animals (rat, mouse, dog) actually are representative test systems for the human body.

Furthermore, toxicological properties are tested in target plants, soil organisms (micro-organisms, earthworms), wild mammals, birds, insects, and surface water organisms (algae, daphniae, fish).

In the case of these non-target organisms the relevant ones can be tested directly and therefore uncertainty does not play a role originating from extrapolation. Nevertheless, the number of well established test species is quite limited because of still unsolved problems with the standardization of the cultivation of other representative organisms.

To discover all the relevant translocation and degradation processes two sets of experiments have to be carried out. In the first tier, radioactively labelled substances are tested in confined laboratory or greenhouse scale models to characterize all the relevant degradation products in plants, laboratory and farm animals, soil, surface water and fish, if necessary during the relevant time period after application until the final fate of the active ingredient and its metabolites. The major objective of these studies is to define the qualitative nature of the residues in the relevant food and feed items as well as in the media (e.g. water, aquatic plants and sediment for fish) via which humans and non-target organisms might come into contact with pesticide residues.

In addition, these studies are used to estimate quantitatively the fate of pesticides such as degradation rate in soil and water and residue level in food items such as plants and lifestock. If these values reflect very clear and positive situations e.g. half-life in soil less than 20 days or no detectable residues above 0.01 mg/kg in food items, they permit a decision not to go on to the next tier of extended testing. However, if this first data set needs further clarification, the next level of studies must be carried out, mostly field studies.

These field studies, form the second tier of tests. In preparing for this analytical methods are developed for the active ingredients and all the relevant metabolites. These methods are used as the safety experts, 'trace analytical glasses' to show all the contaminations of humans and non-target organisms, which may occur in practice, and quantitatively under test conditions in the field studies. These field studies follow good realistic agricultural practice. In order to consider variations in application techniques, weather, soil and cropping scenarios, the worst case situation is tested. When testing leaching to groundwater, sandy soils are used, whilst soils with high clay content are selected to study run-off processes. For applicator exposure the equipment and technique is used which normally shows comparatively high skin or air contamination. In supervised residue trials the highest rate and number of applications as well as the shortest pre-harvest intervals are investigated.

In field studies two additional tests are applied, which are not included in the first level; applicator exposure, and exposure during re-entry into the crop after application, as well as during production and formulation. Furthermore, drift is only measured in field studies. In these situations the active ingredient is the only contaminant which is translocated and it is assumed that no conversion to metabolites occurs during this initial non-intended translocation process.

Test strategies

It is generally accepted that after having generated a solid and complete basic data set a first preliminary risk assessment is made. Whenever this shows a comfortable distance between maximum exposure to residues and non-toxic threshold levels, further testing is not necessary. However, if this first very conservative assessment using a lot of worst case assumptions and extremely unfavourable scenarios demonstrates a critical parameter, a tiered approach of further extended testing in more realistic and specific situations, finally in the actual field, must be applied. The results of the later testings are then the basis of a final evaluation and registration decision.

Validity of data

All standard studies dealing directly or indirectly with risk-related parameters follow test protocols which are approved by national or international scientific bodies. Approvals are issued after the successful execution of inter-laboratory comparative tests. This validation procedure is an obligatory prerequisite for comparable results and decisions.

In addition to this validation measure, the rules of good laboratory practice (GLP) are also applied to all the risk related investigations. These GLP guidelines, issued by different bodies like OECD, US Environmental Protection Agency (EPA), European Community (EC) have improved the quality of safety testing. Additionally, such guidelines guarantee the option to demonstrate the validity of all measurements and allow complete re-evaluations when related findings in other studies require this. Company external and internal inspectors check that the testing facilities and reports plus the corresponding raw data comply with GLP requirements.

It can be concluded that the necessary information for the assessment of risk from pesticide residues is generated when the maximum exposure is measured in realistic worst case conditions for humans and relevant representative non-target organisms, and the non-toxic threshold levels for all the relevant organisms are determined.

Risk assessment and setting of standards for margins of safety

As mentioned earlier, the degree of risk is determined by comparing the maximum exposure of an organism with its own non-toxic threshold exposure level. This approach is rational and scientific and can be understood everywhere in the world, in every agency, and in every research institute. The basis of this assessment is conservative and closely related to actual situations as realistic worst case conditions are investigated for exposure measurements, and the most susceptible organisms are selected to define the borderline between hazard safety, and the non-toxic threshold level. Based on this risk assessment, agencies decide whether the particular pesticide candidate is safe enough and can be licensed to be sold and widely used in agriculture.

The question to arise from this information is, what is safe enough? This becomes more complicated when the certainty and variation of the two parameters of maximal exposure and non-toxic threshold levels are considered. Whatever the test strategy, only a very limited number of situations can be considered and simulated in comparison to the broad variety of scenarios which are relevant in real but good agricultural practice. Realistic worst case conditions are tested in order to simulate the most critical edge of the spectrum.

Furthermore, some assumptions are made which are definitely too conservative in comparison to practice. But these assumptions are clear and simple e.g. that the total area for a crop in a country is treated with a particular pesticide for a whole lifespan of a human. This assumes that an unprocessed crop completely contaminated at the maximal residue level is consumed by every individual human in a country during his whole lifetime. Actually, this is not the case and residue levels go down during normal processing and food preparation. When assessing potential adverse effects on the aquatic environment the maximum exposure via run-off and drift are assumed. This assumption is not true in every case, nor is the high susceptibility of aquatic organisms tested in glass containers in the laboratory, where no adsorption on sediments or suspended sediments occurs.

On the other hand, slightly worse exposure conditions may occur when different influences are considered together under extremely unfavourable conditions leading to even higher exposure than estimated by the risk assessment. Similarly, it might be that even more sensitive organisms are exposed in the environment than those selected for toxicological testing.

When taking all these uncertainties into account, establishing 'margins of safety' between the highest exposure assumed and the lowest non-toxic effect level measured is a widely applied approach. This setting of margins of safety specifically for the precautionary protection of human health is based on political decisions. Thus it is a political decision whether low or higher risks are tolerated in relation to the benefit of the use of a particular pesticide. When very high safety standards for one parameter are demanded, and as a consequence there are bans or severe restrictions on pesticide products with other very positive properties, a drain of resources might be the consequence as well as the loss of problem-solving tools for farmers.

As previously stated, to clarify whether a critical compound is acceptable in its safety aspects, investigations must sometimes be extended to the next level of tests which are closer to actual practice. In field tests, the buffering capacity of an ecosystem in relation to the reduction of a population of one non-target species, and its recovery, can be tested and taken into account when assessing the certainty and variations of the borderline between sustaining effects and effects which are completely reversible. If the safety properties of a pesticide are critical, the benefits of its use and the disadvantages when it is not used must also be taken into account.

Executive information derived from safety testing for the development of safety precautions

All the investigations and their results apply for good agricultural conditions. Consequently, the decision made by an agency to licence a pesticide states:

This plant protection product does not pose any risk to humans or the environment as long as good agricultural practice is applied.

When no precautionary measures are applied, for some products the margins of safety are not large enough for all situations where humans or non-target organisms may be exposed to pesticide residues. In order to increase these margins to acceptable levels, specific warnings and safety measures are defined and given on the product label. For example, specific protective clothing might be necessary to reduce the actual exposure to applicators. Re-entry restrictions might be necessary to protect people during harvest operations. Application during daytime when bees can be contaminated might be prohibited. All these additional safety precautions are communicated on the label. The executive information provided is a result of intensive investigations and extensive reports. The first-aid information and antidotes as well as the guide to the physician given on the label must be considered in the case of an accident.

A pesticide is a safe tool in the hands of the farmer only when it has passed all tests and assessments as acceptably safe and the rules of good agricultural practice are followed, together with the specific instructions on the label.

Information transfer of research results, safety standards and protective measures to retailers and farmers

The whole process from pesticide development to export and use by farmers is a joint effort with shared responsibilities. The definition of a responsible person is one who is competent and who can foresee the consequences of an operation which he is executing. Consequences include benefits as well as risks and potential side-effects. These can only be assessed by a competent person in the country where the pesticide is used. Only he has all the information that is necessary to foresee the outcome. He has access to the data generated by the manufacturer and the information for the scenarios of pesticide application in his country. The manufacturer of a particular pesticide provides specific information about safety precautions and the World Association of Pesticide Manufacturers (GIFAP) provide a large amount of well written and illustrated instructions for safe transport, storage, and application (GIFAP, 1988a) as well as a manual for a course for agrochemical retailers (GIFAP, 1988b). These are supplemented with video tapes.

In addition the manufacturers have developed and selected materials which show the best protection against unintended contamination such as overalls, rubber gloves, trousers, boots, faceshields for full face protection, safety glasses, face masks to protect against dust, and aprons for extra protection when mixing and loading pesticides. All this information can be supplied by the manufacturer and can be adapted in the importing country according to the circumstances of pesticide application.

As it is said in the FAO Code of Conduct, Article 1 – objectives of the code (paragraph 1.3):

The Code addresses the need for cooperative effort between governments of exporting and importing countries to promote practices which ensure efficient and safe use, while minimizing health and environmental concerns due to improper handling or use.

These voluntary standards endeavour to ensure that there is no gap of responsibilities between the manufacturing country and the country where a pesticide is used.

Communicating risk and safety to the public

Again, the Code of Conduct gives guidance in this issue in paragraph 1.4 of the objectives:

The entities which are addressed by this Code include: international organizations; governments of exporting and importing countries; industry, including manufacturers, trade associations, formulators and distributors; users; and public sector organizations such as environmental groups, consumer groups, and trade unions.

The 'entities' addressed, in the sense of this Code, means all those who take a partial responsibility, including the public sector organizations. Environmental and consumer groups are responsible for raising questions from their point of view. These must be answered by the manufacturers and registration authorities both of whom have the complete set of data concerning all safety factors. If necessary, additional investigations have to be carried out in order to clarify public concern.
However, all other groups, including the public sector organizations, which do not have access to the complete data sets, are not competent to give final answers or assess adequately one or a group of pesticides. The whole issue of the benefit–risk problem is too complex to be covered in one public statement or article in a newspaper. Unbalanced and incorrect statements can cause damage to a tool which enables farmers to produce more food. The effects of such questions can be severe. Political decisions may be unreasonably influenced resulting in very expensive but nevertheless useless studies.

Money spent on futile studies is therefore not available for investigating new and better products nor investigating the reduction of objectively higher risks. To achieve the objective of making the most effective use of research data and resources, a fair and open communication of risk issues between the experts in the industry and the agencies on the one hand, and the public on the other hand, must be developed. The experts have to improve their presentation of research data, and the public sector organizations should accept and distribute 'good news' about the safe use of pesticides.

References

GIFAP (1983) *Guidelines for the safe and effective use of pesticides.* GIFAP, Brussels.

GIFAP (1988a) *GIFAP Trainers Manual, Course for Farm Trainers.* GIFAP Brussels.

GIFAP (1988b) *GIFAP Trainers Manual, Course for Agrochemical Retailers.* GIFAP, Brussels.

GIFAP (1989) *Guidelines for personal protection when using pesticides in hot climates.* GIFAP, Brussels.

Chapter 6
Herbicide Resistance

F.W.G. Baker

Scientific Secretary, CASAFA

Introduction

Blackman (1950) had already pointed out the similarities between the mass selection for a given character in crops and the selection of weeds after continuous herbicide treatment in fields when the development of resistance to herbicides began to be noticed in the 1970s. Since then repeated annual applications of chemically similar herbicides, but especially of triazine compounds, particularly on crop monoculture, perennial plantations or on wasteland, have led to the development of resistance in weed populations becoming a global phenomenon.

Resistance and cross resistance

Resistance

Since Ryan (1970) reported the resistance of common groundsel, *Senecio vulgaris,* to simazine and atrazine there have been many reports of the development of resistance to various herbicides in a widening range of weedy species. However, the best-studied example of resistance of weedy biotypes is that of triazine compounds. For example, LeBaron and McFarland (1990) list more than 50 species that have resistance to herbicides, with at least 34 of these resistant to triazine; Darmency and Gasquez (1990a) list 19 grasses and 48 broadleaf species with biotypes resistant to herbicides. LeBaron and Gressel (1982) in the conclusion to *Herbicide Resistance in Plants* wrote: 'there is no doubt that there will be many new reports of triazine resistant weeds in the near future. Undoubtedly the list of resistant biotypes may become longer for other herbicides as well'. The list has certainly become longer, but increased awareness of the problems and the introduction of certain countermeasures seem to have reduced the number of cases of new herbicide resistant weeds reported each year.

LeBaron and McFarland (1990, Table VII) list the number of states, provinces and countries reporting the first appearance of herbicide resistant biotypes by year: that for triazine resistance peaks in 1984 (1988 being the last year reported). In 1974, Peabody drew attention to the occurrence of a population of *Amaranthus* spp. in western Washington which was resistant to high levels of one of the triazines, atrazine. At present there are more than 50 weeds with resistance to triazine compounds, almost five times that for any other herbicide group. Triazines are residual herbicides with soil activity that usually persists in the soil during the season of application and sometimes into the second season. This is of particular importance for the length of selection pressure and this, the long period of use and the over-confidence in the efficacity of triazines may be the main reasons why so many weeds have developed resistance to triazines. Some other herbicides, such as chlorsulfuron and imidazolinones, also persist for similar periods of time in the soil but have been introduced more recently and the number of weed biotypes resistant to them is still small.

Cross resistance

Oettmeier *et al.* (1982) indicated that in the case of atrazine-resistant biotypes cross resistance occurs to all compounds of the triazine group. The first report of a weed species that exhibited extensive cross resistance to chemically dissimilar herbicides and to herbicides with different modes of action is that for annual ryegrass, *Lolium rigidum*, in Australia by Heap and Knight (1986). Also from Australia, Powles and Howat (1990) indicate that from two to three years use of eight triazines, seven substituted ureas, amitrole and metribuzin may result in failure to control annual ryegrass because of further development of cross resistance. Cross resistance has also been observed in *Alopecurus myosuroides* in central and eastern England (Clarke and Moss, 1989), Germany and Israel.

Mechanisms of resistance

There appear to be two main mechanisms of resistance: a) alteration of the target site of herbicide action, for example the binding site of electron transport inhibitors of photosystem II; and b) detoxification or metabolism of the herbicide in the plant, for example glutathione-S-transferase is not only associated with the metabolism of triazine in maize but also with the detoxification of EPTC.

Alteration of target site

Chloroplastic resistance to triazines, which was first reported in the early 1970s, was the first observed resistance to herbicides. The resistant populations usually appear after continuous use of the same herbicide for several years. (In the first cases observed this was after about five to ten years of triazine use but resistance to some of the more recently introduced herbicides has appeared after shorter periods of use.)

The mechanism of herbicide resistance to sulfonylurea and other types of herbicides with acetohydroxyacetate synthase inhibitors (AHAS) appears to result from an alteration in the gene coding for AHAS. As a consequence, insensitive AHAS enzymes are produced. Darmency and Gasquez (1990b) state that:

> all triazine resistant biotypes in which chloroplast DNA has been analysed are mutated on the psbA gene on the chloroplast DNA. This mutation corresponds to the change of one aminoacid at position 264 of the psbA gene product, the 32 KD protein. The fact that the mutation occurs on the psbA gene is not surprising because atrazine is a competitor of quinones binding to the 32 KD protein.

It seems probable that all weed species possess some ability to develop resistance to herbicides, such as sulfonylurea, that use AHAS inhibitors. Some weeds that are cross resistant to several herbicides have more than one mechanism. For example, most triazine-resistant biotypes have alterations in the triazine-binding properties in their chloroplasts. Isolated chloroplasts from resistant weed biotypes have been shown to be unaffected, when assayed *in vitro*, by atrazine and are able to withstand doses 1000 times those that block the target site in the chloroplast in sensitive biotypes (Darmency and Gasquez, 1990).

Vaughn and Vaughn (1990) have carried out studies of dinitroaniline-resistant biotypes of *Eleusine indica*, one of which is highly resistant and another has intermediate resistance. They suggest that because the highly resistant biotype is hypersensitive to the microtubule-stabilizing agent taxol the resistance is likely to result from its having hyperstabilized microtubules.

With regard to the resistance of *Alopecurus myosuroides* to substituted ureas, Chauvel and Gasquez (1990) indicate that this is due to a mutation of the target of herbicide action in the chloroplast which has become insensitive to the herbicide. They indicate that the mechanism of herbicide detoxification in *A. myosuroides* is 'polygenic, nuclearly inherited and, according to the biology of the species, may reduce the spread of resistance'.

Detoxification or metabolism of the herbicide

Some weeds – and crops – have the ability to bind the herbicide outside of the chloroplasts or to metabolize the active chemical substance before it can have a lethal effect. Clay (1989) indicates that for hop-garden biotypes of *Poa annua*, the mechanism of resistance may be different from that normally found in triazine-resistant biotypes of this and other species. He suggests that the resistance may be due to a detoxification mechanism reducing the amount of herbicide reaching the site of action, comparable to that found with *Alopecurus myosuroides* resistant to chloroturon and other similar herbicides.

For *Lolium rigidum* Powles and Howat (1990) suggest that it seems possible that cross resistance is due to an enhanced ability to metabolize certain herbicides by elevated activity of the polysubstrate monooxygenase suite of enzymes. Powles *et al.* (1990) indicate that the mixed function oxidase inhibitors aminobenzotriazole and piperonyl butoxide can synergize herbicide activity in resistant *L. rigidum* growing in a hydroponic system and suggest that at least one aspect of cross resistance in *L. rigidum* may be related to enhanced metabolic activity of mixed function oxidases.

With regard to resistance to paraquat in *Conyza bonariensis*, the detoxification appears to be correlated to the activities of three enzymes, superoxide dismutase, ascorbate peroxidase and glutathione reductase, which enable the chloroplasts at photosystem I to remove hydrogen peroxide and superoxide formed after the paraquat treatment (Shaaltiel and Gressel, 1986).

Upon failure of one herbicide, a switch to one or more alternative herbicides may, or may not, control the weed. For example, Powles and Howat (1990) show that for *Lolium rigidum*, resistant to diclofop-methyl, a switch to one or more alternative herbicides may, or may not, control the weed but that two to three years uninterrupted use of alternative herbicides may result in failure due to the further extension of cross resistance.

Herbicide resistant crops

In addition to the strategies suggested for the resolution of the problem of herbicide resistance using different cultivation practices and patterns of different herbicide use, the possibility also exists of modifying the crop plants to reduce susceptibility or to enhance resistance to herbicides.

Classical breeding methods and genetic engineering have already demonstrated their potential to provide this. Classical breeding methods were first used to introduce atrazine resistance from *Brassica campestris* to *B. napus* (Beversdorff and Kott, 1987). Genetic engineering techniques, such as protoplast fusion, are being used to transfer resistance from weeds to related crops when they do not interbreed, for example from *Solanum nigrum* to *S. tuberosum*.

Work is currently in progress to produce a range of crop cultivars, including cotton, maize, soyabean, sunflower, tobacco, tomato, etc. with resistance to several herbicides.

Newhouse *et al.* (1990) indicate: 'the project to develop corn hybrids resistant to imidazolinone herbicides has been very successful' with the development of a hybrid corn – herbicide package where the 'commercial corn hybrids resistant to imidazolinone herbicides will be the first major biotechnology product in cereals to enter the market place'.

The development and large-scale introduction of herbicide resistant crops will need careful and long-term study for several reasons. First is the possibility of the unwanted spread of genes with herbicide resistance, from crops to weedy relatives. (Particular care will be needed to ensure that seeds from herbicide resistant crops are not exported, for example as food aid, to developing countries where the crop would probably be grown without the herbicide but in the presence of wild relatives.) Secondly, it will be necessary to ensure that the introduced genes do not result in the development of toxic compounds in the crops. (The classic case of the introduction of pest resistance into *S. tuberosum* using wild genes led to the production of toxic concentrations of solanine: this was discovered shortly before the intended release of the new, resistant, but toxic, variety on the market.) Thirdly is the possibility of the resistant crop self-seeding and causing problems in crop rotations. Finally, as there is already a tendency to reduce the number of available herbicides, the introduction of a still smaller number of crop-related herbicides could increase selection pressures leading to the further development of resistant weed biotypes.

Herbicide resistance in the tropics

At present there are only a few tropical or semi-tropical countries that have reported herbicide resistant weed biotypes, for example Colombia with propanil resistant *Echinochloa crus-galli*, but there have been no cases reported of triazine-resistant weeds in low input systems, probably because of: i) costs; ii) different management practices; and iii) fewer informed observations. Collins (Chapter 4, this volume) suggested that cases of resistance in low input tropical agriculture are likely to be less than in the temperate regions because of the time taken for resistance to develop.

There are also indications that there is increased heat sensitivity of the photosynthetic apparatus in certain triazine-resistant weed biotypes. It seems likely, however, that in low input tropical agriculture, factors of economy will lead to the use of the cheapest, and therefore probably the same, herbicide. Particular attention will therefore have to be paid to ensure that the timing, rates and methods of application of herbicides are most effective and least likely to lead to the development of resistance.

If one looks at the list of the world's worst weeds (Holm *et al.*, 1977) it is interesting to note that only seven of the 20 worst weeds have been reported to have resistance and four of these are grasses. From this one might deduce that the tropical grassy weed that is likely to cause the greatest problems is *Echinochloa crus-galli*, which has already shown resistance to triazine and propanil.

Conclusions

Biotypes that have developed resistance to a herbicide can usually be controlled by alternative, chemically different herbicides and do not cause particular problems. However, biotypes that have developed cross resistance can prove difficult to control.

Although the use of fungal herbicides (Evans, Chapter 3, this volume) is still in its early stages, such methods may become of particular importance in the control of cross resistant weed biotypes.

More work is needed to further develop improved types of integrated weed management strategies (Aeschlimann, Chapter 7, this volume) and to prevent, or at least delay, the emergence of herbicide resistant and cross resistant weed biotypes. In particular more studies are needed into the effects of agronomic conditions and of the timing and methods of application of herbicides. Although many groups of scientists are carrying out research into the mechanisms of cross resistance and into the development of effective methods for the control of cross resistant biotypes, up to the present integrated strategies, of the type provided by the Australian Herbicide Resistance Action Committee (AHRAC) Figures 6.1 and 6.2, seem to provide the best chance of success. The strategy is based on an integrated weed management system that includes crop rotation, herbicide rotation and various cultivation practices, including burning. LeBaron and McFarland (1990) have also given a list of 13 changes or strategy rules they believe will be required to protect, preserve or prolong the use of the newer herbicides with a single site of action and therefore at high risk for developing resistance. (Other authors, for example Clarke and Moss (1989), Gressel and Segel (1990), Maillet (1991) have also put forward suggestions for control that include herbicides, crop rotations and different cultivation practices). With regard to cultivation practices Putwain and Collin (1989)

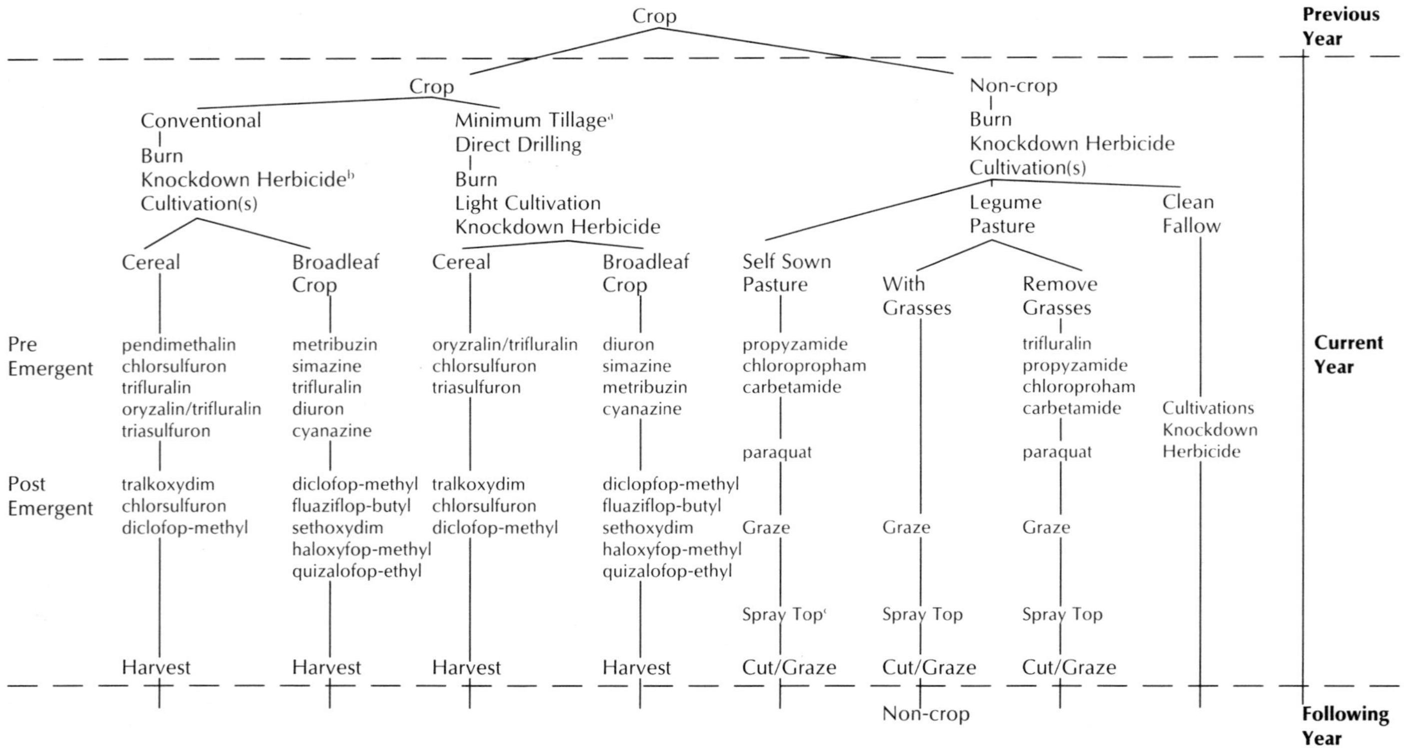

Fig. 6.1 Integrated strategy for annual ryegrass control, to minimize resistance.

[a]Minimum Tillage/Direct Drilling – systems in which mechanical weed control by cultivation is reduced or replaced by knockdown herbicides.
[b]Knockdown Herbicides – the use of glyphosate or paraquat to control seedling weeds.
[c]Spray Top – the use of low rates of knockdown herbicides, glyphosate and paraquat to reduce seed set.

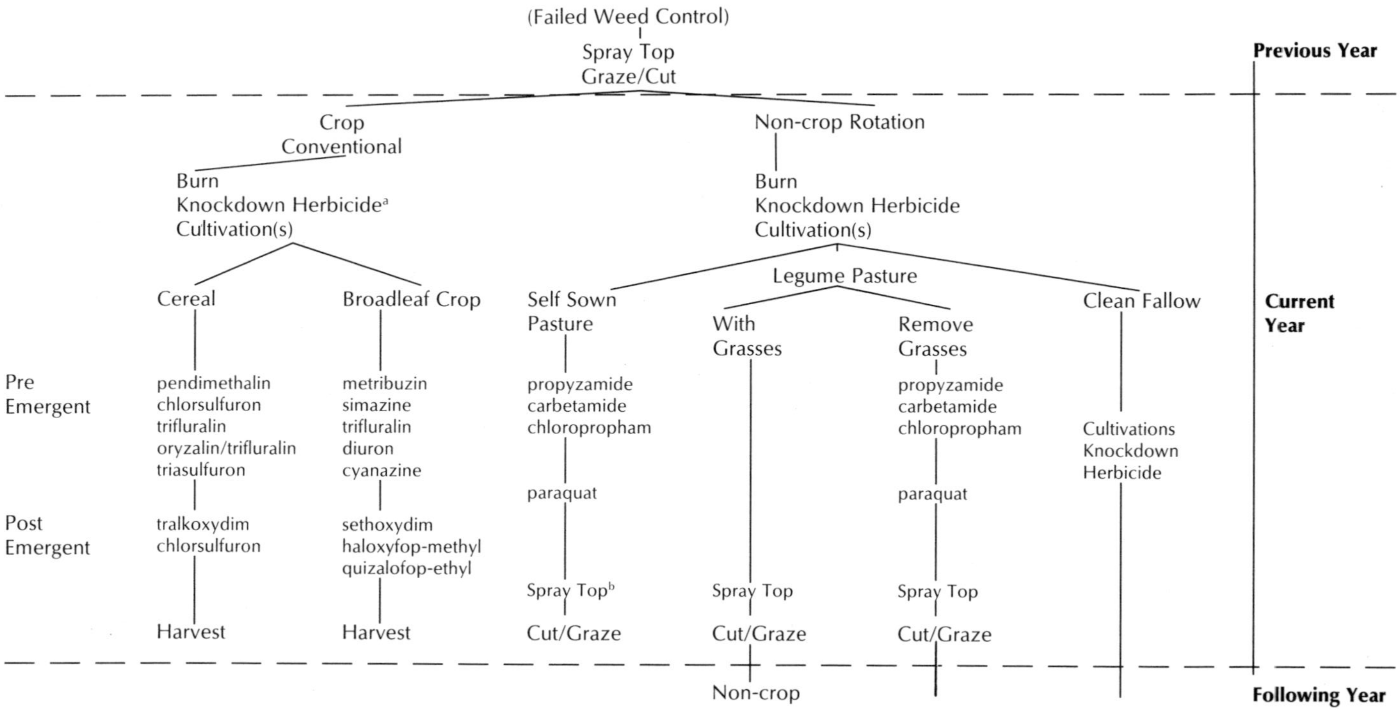

Fig. 6.2 Integrated strategy for annual ryegrass control, to manage confirmed resistance.

[a]Knockdown Herbicides – the use of glyphosate or paraquat to control seedling weeds.
[b]Spray Top – the use of low rates of the knockdown herbicides, glyphosate and paraquat to reduce seed set.

have drawn attention to the potential problems of persistent buried dormant seed populations of some resistant biotypes of weeds.

The genetics of the world's worst weeds, the effects of the mutations that lead to herbicide resistance and the effects of such mutations on the competitiveness of resistant weed biotypes also require greater research efforts.

It seems probable that the greatest problems will be encountered where the use of herbicides with high selection pressure rates have resulted in resistant mutants that are both competitive and fit.

As the use of herbicides increases in developing countries the distributors of herbicides should be made aware of the problems created by the development of resistance and every effort made to ensure that extension workers and farmers are provided with information and appropriate strategies to prevent or delay extending resistance to a still wider range of weeds.

Acknowledgements

I thank J. Gasquez for useful discussions, valuable comments and for the use of his library, the INRA Laboratoire de Malherbologie, Dijon, for the use of its library and J.J.S. van Rensen for valuable comments.

References

Beversdorf, W.D. and Kott, L.S. (1987) Development of triazine resistance in crops by classical plant breeding. *Weed Science* 35 (Supplement 1), 9–11.

Blackman, G.E. (1950) *Journal of the Royal Society of Arts*, 499–517.

Cavallaro, R. and Noyé, G. (eds) (1989) *Importance and Perspectives on Herbicide-resistant Weeds.* CEC Publications, Luxembourg.

Chauvel, B. and Gasquez, J. (1990) Résistance aux urées substituées: le cas du Vulpin. *La Défense des Végétaux* 263, 22–7.

Clarke, J.H. and Moss, S.R. (1989) The distribution and control of herbicide resistant *Alopecurus myosuroides* Huds. (black-grass) in central and eastern England. *Proceedings of the Brighton Crop Conference, Weeds 1989*, 301–8.

Clay, D.V. (1989) New developments in triazine and paraquat resistance and coresistance in weed species in England. *Proceedings of the Brighton Crop Conference, Weeds 1989*, 317–24.

Darmency, H. and Gasquez, J. (1990a) Résistance aux herbicides chez les mauvaises herbes. *Agronomie* 6, 457–72.

Darmency, H. and Gasquez, J. (1990b) The fate of herbicide resistance genes in weeds. In: Green, M.B., Moberg, W.K. and LeBaron, H.M. (eds) *Managing Resistance to Agrochemicals*, ACS Symposium Series, Washington.

Green, M.B., Moberg, W.K. and LeBaron, H.M. (1990) (eds) *Managing Resistance to Agrochemicals*, ACS Symposium Series, Washington.

Gressel, J. and Segel, L.A. (1990) Herbicide rotations and mixtures. In: Green, M.B., Moberg, W.K. and LeBaron, H.M. (eds), Managing Resistance to Agrochemicals ACS Symposium Series, Washington, pp 430–51.

Heap, J. and Knight, R. (1986) The occurrence of herbicide cross-resistance in a population of annual ryegrass, *Lolium rigidum*, resistant to diclofop-methyl. *Australian Journal of Agricultural Science* 48, 56–157.

Holm, L.G., Plucknett, D.L., Pancho, J.V. and Herberger, J.P. (1977) *The World's Worst Weeds: Distribution and Biology.* The University Press of Hawaii, Honolulu, 609pp.

LeBaron, H.M. and Gressel, J. (eds) (1982) *Herbicide Resistance in Plants.* Wiley-Interscience, New York.

LeBaron, H.M. and McFarland, J. (1990) In: Green, M.B., Moberg, W.R. and LeBaron, H.M. (eds) *Managing Resistance to Agrochemicals.* ACS Symposium Series, Washington, pp336–52.

Newhouse, K.E., Shaner, D.L., Wang, T. and Fincher, R. (1990) Genetic modification of crop responses to imidazolinone herbicides. In: Green, M.B., Moberg, W.K. and LeBaron, H.M. (eds) *Managing Resistance to Agrochemicals.* ACS Symposium Series, Washington, pp474–81.

Oettmeier, W., Masson, K., Fedtke, C., Konze, J. and Schmidt, R.R. (1982) Effect of different photosystem II inhibitors on chloroplasts isolated from species either susceptible or resistant toward s-triazine herbicides. *Pesticide Biochemistry and Physiology* 357–67.

Peabody, D. (1974) Herbicide tolerant weeds appear in western Washington. *Weeds Today* 5, 14.

Powles, S.B. and Howat, P.D. (1990) Herbicide resistant weeds in Australia, *Weed Technology* 4, 178–85.

Powles, S.B., Holtum, J.A.M., Matthews. J.M. and Liljegren, D. (1990) Herbicide cross-resistance in annual ryegrass (*Lolium rigidum* Gaud.): the search for a mechanism. In: Green, M.B., Moberg, W.K. and LeBaron, H.M. (eds) *Managing Resistance to Agrochemicals.* ACS Symposium Series, Washington, pp394–405.

Putwain, P.D. and Collin, H.A. (1989) Mechanisms involved in the evolution of herbicide resistance. In: Dodge, A.D. (ed.) *Herbicides and Plant Metabolism.* SEB Seminar Series 38, CUP, Cambridge.

Ryan, G.F. (1970) Resistance of common groundsel to simazine and triazine. *Weed Science* 18, 614–16.

Shaaltiel, Y. and Gressel, J. (1986) Multienzyme oxygen radical detoxifying system correlated with paraquat resistance in *Conyza bonariensis. Pesticide Biochemistry and Physiology* 26, 22–8.

Van Oorschot, J.L.P. (1989) Some consequences of growing herbicide-resistant crops for the development of resistance in weeds. In: Cavallaro, R. and Noyé, G. (eds) *Importance and Perspectives on Herbicide-resistant Weeds.* CGC Publications, Luxembourg.

Chapter 7
Towards Integrated Weed Management

J.-P. Aeschlimann

Secretary General, International Organization for Biological Control, CSIRO Biological Control Unit, 335 avenue Paul Parguel, 34090 Montpellier, France

Summary

Integrated pest management (IPM) is basically the use of all available and compatible techniques and methods to maintain populations of harmful organisms at levels below those causing economic damage. IPM aims first to reduce the inputs of non-renewable resources (chemical substances in particular) by maximizing the impact of the pests' natural enemies. IPM is hence a dynamic concept that involves a tremendous accumulation of ecological, economic and toxicological knowledge. It is certainly fair to say that we are still far from the point at which this strategy could be used by farmers to control adequately plants considered as noxious, especially in the tropics. To meet this challenge, research will have to focus on developing the understanding of the critical periods for the different crops concerned, on defining injury levels for the various undesired plant species and their associations and on finding natural enemies of weeds suitable for use as an alternative to herbicides (inoculative or inundative applications).

Introduction

The need for developing an integrated approach to weed control was already clearly recognized almost 20 years ago by a number of research scientists (see Fryer and Matsunaka, 1977). In the meantime, however, the situation for the control of weeds has seriously deteriorated; weed scientists nowadays being confronted with a dramatic and widespread emergence of weeds that are resistant to all currently used herbicides (cf. Aeschlimann, 1986; Pasteur, 1989). To date, about 50 undesired plant species have been found to be resistant to all triazine compounds and the genetic mechanisms involved in the process have been investigated and described in some detail (Gasquez *et al.*, 1989; Gasquez and Darmency, 1990). Furthermore, the persistent application of the same chemical herbicides, and routinely repeated cultivation practices lead to modifications in weed communities; in particular to a decreasing occurrence of the short-lived species, which are gradually replaced by perennial, herbicide-resistant, weeds. Finally, authorities, farmers and consumers are becoming increasingly aware, in the industrialized world at least, of the dangers that highly residual and toxic herbicides represent for the environment, and especially for the supply of drinking water.

IPM: definition and implications

Although it is implicitly included in the concept of integrated pest management, a proper 'integrated weed management' scheme has not yet been developed. Even extremely thorough and recent textbooks on IPM, which unequivocally consider pests to be all noxious animals and plants, do not deal at all with the practical problems posed by the presence of weeds or deleterious micro-organisms in agro-ecosystems (cf. for instance Horn, 1988; Riba and Silvy, 1989). These, however, are examples of authors, who respected the original definition of IPM, which, in the sense that has been widely accepted among scientists, is a strategy '... which uses all economically, ecologically, and toxicologically justifiable methods to keep pests below the economic thresholds, with the emphasis on the conscious utilization of natural control factors ...' as was summarized by Steiner (1988).

Many so-called plant protection specialists, on the other hand, claim to be applying IPM in a number of cropping systems simply because selected seed material is planted, on which cultivation practices alternate with pesticide applications, the chemical substances being chosen to avoid build-up of resistance in the target pests. This is a distorted conception of IPM, which ought to be firmly rejected.

The IPM strategy was originally designed by entomologists with the declared aim of reducing the use of non-renewable inputs (mainly acaricides and insecticides) to a minimum, ideally to zero. An in-depth understanding of all parameters likely to play a part during the whole growing season was necessary to construct a comprehensive model for any production process. An average of 10 to 20 years was needed to gather enough reliable data to develop an acceptable, working model for just one individual cropping system. In addition to this tremendous accumulation of knowledge, IPM also requires high flexibility so that the constantly changing economic constraints can be considered and incorporated and the appropriate decisions can be made on time; that is to say according to the population numbers of the different living species actually present in the ecosystem.

The nucleus of the IPM strategy is constituted by the 'threshold of economic injury', which has to be adopted and adapted by each practitioner (cf. Horn, 1988), to decide whether control measures are required or not and of what type.

Population dynamics relationships

The basic idea of IPM is to proceed with a curative treatment only when the economic damage caused by the occurrence of a noxious organism in the field will be equal or superior in terms of yield loss to the total expenses incurred by that treatment. This approach has two main practical implications: (i) the model developed over the years must be trusted to provide an accurate and reliable forecast of the loss ahead of action; and (ii) the level of efficiency of the control method ought to be strictly guaranteed (need for quality control).

Entomologists and acarologists have learned how to monitor the population dynamics of the various phytophages involved, as well as that of their natural enemies, and, as a consequence, to apply efficient and safe control procedures. In the case of integrated weed management, the situation is much more complicated, plant communities being usually composed of a whole series of detrimental species which should first be ranked according to their competitiveness. For practical purposes, the number of individuals of the predominant weed species has often been considered in isolation, thus grossly exaggerating the weed's incidence (1% yield reduction due to each *Avena fatua* L. plant per m^2 for instance), the economic threshold (2.5 *A. fatua* individuals per m^2 soil surface already), and hence the benefits likely to arise from chemical control (total cost for a single herbicide application against *A. fatua* not exceeding 0.5–1.0% of the gross product value according to Cussans, 1988).

Apart from the economic threshold for weeds, which in most crop situations remains to be determined, the 'critical period' is of foremost importance. This is the time during which weeds have to be removed to provide the cultivated plant with the opportunity to express its full yield potential.

Critical periods

Evidence is now available for several cultivated plants that the plots only need to be weedfree during a very limited number of weeks each season. The critical period for maize lasts approximately from the 2–3 leaves stage until the plants are about 30–40 cm high, which does not represent more than one month under European conditions, i.e. a slight proportion of the entire growing season (Ammon, 1988). In the case of sugarbeet, weeding before the 3rd, and after the 8th week following seedling emergence does not increase the total yield (Ammon, 1988). Priority should therefore be given to a better understanding and definition of critical periods for the various production systems concerned.

The determination of reliable critical periods, in addition to the use of realistic levels of economic injury might already contribute to reduce considerably the number of herbicide applications per year. If IPM has to become a viable strategy against weeds, however, efficient substitutes to the chemical substances will also have to be discovered and made available at a reasonable price to the farmers.

Search for alternatives to herbicides

Based on a survey of recent literature, there are only two major examples of integrated weed management strategy that can be classified as lasting successes in relation to accidentally introduced important terrestrial weeds, i.e. musk thistle, *Carduus nutans* L. (Compositae), in North America (cf. for instance Kok and Pienkowski, 1985), and skeleton weed, *Chondrilla juncea* L. (Compositae), in southeast Australia (Groves and Cullen, 1981). In both cases, substantial and sustained control has been achieved by the well adjusted combination of deliberately introduced natural enemies (phytophagous insects and weed pathogens), efficient cultivation practices, and the timely introduction of useful plants (Gramineae and Papilionaceae) able to compete with the undesired species in the community.

Many promising avenues in terms of biological control of adventitious plants remain to be explored, as a study of the catalogue by Julien (1987) would suggest. For example, phytophagous insects that feed on weeds may in fact be much more restricted in the choice of their host plants than was assumed previously: for instance, a biochemical analysis carried out by Klein (1991) has demonstrated that the weevil *Rhinocyllus conicus* Fröhlich, which is playing a key role against nodding thistles, was actually represented by a whole range of sometimes very different populations, associated with various thistle species in the Palaearctic region. An unexpectedly high and stable level of specificity has also been found among particularly virulent plant pathogens, like the rust fungus *Puccinia chondrillina* Bubak & Syd., each strain of which generally affected just one of the many clones of skeleton weed identified so far (Chaboudez, 1989). Consequently, as Evans (Chapter 3, this volume) has exposed in some detail, mycoherbicides could well offer an ideal solution for a number of situations.

Investigations with a view to developing a strategy of integrated management against adventitious plants occurring in vines and soft fruits (Cavalloro and Robinson, 1987), or in vegetable production (Cavalloro and El Titi, 1988) are under way, mainly in industrialized countries. In the tropics and subtropics, the problems posed by perennial grasses certainly call for immediate attention, and there is little doubt that expertise already available elsewhere could be extremely helpful in this region, provided it is accompanied by adequate funding of research and field implementation.

Acknowledgements

The author is grateful to D. Briese (Montpellier) for valuable comments during the early stages of the preparation of this manuscript.

References

Aeschlimann, J.P. (1986) Postface. In: Van den Bosch, R. and Aeschlimann, J.P. (eds) *L'Engrenage des Pesticides.* Payot, Lausanne, pp223–51.

Ammon, H.U. (1988) Do actual thresholds favour the build-up of herbicide resistance in weed populations? In: Cavalloro, R. and El Titi, A. (eds) *Weed Control in Vegetable Production.* Balkema, Rotterdam, pp281–9.

Cavalloro, R. and El Titi, A. (eds) (1988) *Weed Control in Vegetable Production.* Balkema, Rotterdam, 303pp.

Cavalloro, R. and Robinson, D.W. (eds) (1987) *Weed Control on Vine and Soft Fruits.* Balkema, Rotterdam, 165pp.

Chaboudez, P. (1989) Modes de reproduction et variabilité génétique des populations de *Chondrilla juncea* L.: implications dans la lutte microbiologique contre cette mauvaise herbe. *Thèse de Doctorat,* USTL, Montpellier, 161 + 4pp.

Cussans, G.W. (1988) The economics of weed control in vegetable research. In: Cavalloro, R. and El Titi, A. (eds) *Weed Control in Vegetable Production,* Balkema, Rotterdam, pp99–105.

Fryer, J.D. and Matsunaka, S. (eds) (1977) *Integrated Control of Weeds.* University of Tokyo Press, 262pp.

Gasquez, J. and Darmency, H. (1990) Herbicide resistance in France, In: Cavalloro, R. and Noyé, G. (eds) *Importance and Perspectives on Herbicide-resistant Weeds.* Office of Official Publications of the EC Luxembourg, pp81–94.

Gasquez, J., Darmency, H. and Dron, M. (1989) L'efficacité limitée des herbicides. *Courrier du CNRS* 72, 76.

Groves, R.H. and Cullen, J.M. (1981) *Chondrilla juncea*: the ecological control of a weed. In: Kitching, R.L. and Jones, R.E. (eds) *The Ecology of Pests. Some Australian Case Histories.* CSIRO, Melbourne, pp7–17.

Horn, D.J. (1988) *Ecological Approach to Pest Management.* Guilford, New York, 285pp.

Julien, M.H. (1987) *Biological Control of Weeds. A World Catalogue of Agents and their Target Weeds.* 2nd edn. CAB International, Wallingford, Oxfordshire, 144pp.

Klein, M. (1991) *Populationsbiologische Untersuchungen an* Rhinocyllus conicus *Fröhlich (Col.; Curculienidae*). Maraun, Frankfurt, 91 + 22pp.

Kok, L.T. and Pienkowski, R.L. (1985) Biological control of musk thistle by *Rhinocyllus conicus* (*Coleoptera: Curculionidae*) in Virginia from 1969 to 1980. In: Delfosse, E.S. (ed.) *Proceedings of the VI International Symposium on Biological Control of Weeds.* Agriculture Canada, Ottawa, pp433–8.

Pasteur, N. (1989) La résistance croissante aux pesticides. *Courrier du CNRS* 72, 74–5.

Riba, G. and Silvy, C. (1989) *Combattre les ravageurs des cultures. Enjeux et perspectives.* INRA, Paris, 230pp.

Steiner, J. (1988) Integrated plant protection – principle and application. In: Cavalloro, R. and Way, M.J. (eds) *Integrated Crop Protection: From Principles to Practical Implementation.* EC Office, Luxembourg, pp29–32.

Chapter 8
Control of Grassy Weeds in Tropical Cereals

J. Maillet

ENSA Biologie végétale, 34060 Montpellier, France

Summary

Grassy weeds are well represented in tropical weed communities; their biological and ecological diversity allow them to appear in most agro-ecosystems. Their similarities with cereals increase their potential noxious effects in these crops. In tropical areas, Pooideae (like *Avena fatua* or *Phalaris minor*) have been introduced with wheat; they occur mostly in the highlands. Panicoideae and Chloridoideae are better adapted to tropical conditions; they represent more than 15% of species richness in tropical cereals.

Grassy weeds seem to be much more harmful under intensive agriculture: fertilization and herbicides may induce shifts in weedy communities in favour of some troublesome Poaceae such as *Rottboellia cochinchinensis*, *Eleusine indica*, *Digitaria ciliaris* or *Echinochloa colona* among annuals, or *Cynodon dactylon* or *Imperata cylindrica* among perennials.

In small-farmer agriculture, grassy weed control must be included in a global weed management plan to avoid critical weed shifts. Chemical weeding, though very promising in experimental trials, is not necessarily the only way; but improvement of other practices requires more biological knowledge. Research priorities among alternative weeding processes should be assigned according to regional particularities, to economical and sociological parameters as well as technical solutions.

Introduction

According to Wellhausen in 1970, at that time one half of the world's population depended on a subsistence, mainly traditional type of agriculture that occupied 40% of arable lands. In the tropics, subsistence agriculture is practised by small farmers on limited areas (1–5 ha), most often with multi-cropping systems, little or no artificial fertilizers or other chemicals, large periods of fallow resulting in itinerant agriculture with slash-and-burn practices (Akobundu, 1982).

Itinerant production is often explained as the consequence of low fertility of tropical soils and rapid development of weedy communities which cannot be controlled by hand-weeding after a few years (Leger, 1990; Marnotte, 1984; Michieka, 1988). The limited extension of cultivated area by one farmer is also related to the amount of work due to traditional weeding in the first weeks of crop establishment (Montillet, 1988). Miller (1982) estimates that in Africa weeding may represent 20–80% of the total labour time in a small-farm cropping system. In Ethiopia, 30–40% of total labour requirement for maize production goes into weed control (Koch *et al.*, 1982).

When periods of fallow are sufficient in length, weed problems in traditional cropping systems are generally 'within the farmer's resource of labour capabilities' (Akobundu, 1980), even if weeding does not appear 'perfect' to those wishing to introduce new technologies (Zimdahl, 1984). In this case, weeds may also play a useful role as forage, medicinal plants, mulch and so on (Chacon and Gliessman, 1982).

However, for several years, increasing population pressures in all countries of tropical areas have led to reduced fallow periods (Akobundu, 1982; Leger, 1990; Traore, 1987), consequently problems of non-restoration of organic matter and nutrients to the soil, erosion and leaching (Swamy and Ramakrishnan, 1988) and increase of weed populations have appeared. Crop yield may decline very quickly if there is no change in agronomic practices.

These problems concern mainly small land-holders and subsistence crops such as cereals which represent a dominant component of human consumption in tropical countries. With permanent settlement, an improvement of cultural practices is necessary at least to maintain food production, and if possible to increase it. Weeding appears to be one of the key improvements (Matthews, 1984; Koch *et al.*, 1982).

Numerous experiments at several research stations have shown that it is possible to control most of the weeds in maize, sorghum and millet (Chabrier, 1986; Deat, 1986; Fussell *et al.*, 1987; Marnotte and Tehia, 1983; Schweizer *et al.*, 1989). But generally, this is possible only in an intensified cropping system using herbicides, fertilization, minimum tillage or mechanical ploughing, and better crop varieties.

Often such methods cannot be applied easily at the farmer's level, as small farmers can not afford resources necessary for the new technologies, or these technologies are not suited to the particular farming system (multi-cropping for example) or specific weed management needs (Maillet, 1991;

Zimdahl, 1984). In some cases, particularly with grassy weeds in cereal crops, weed shifts induced by intensification result in greater competition effects (Fontanel, 1987a; Fisher *et al.*, 1985; Gill, 1986; Schweizer *et al.*, 1989).

There is therefore a need, in order to propose suitable integrated weed management, to analyse every alternative weeding process in relation to economical and sociological parameters in a regional context. Biological and ecological knowledge of weed populations and communities is necessary if weed scientists want to offer long-term solutions.

Tropical cereals

The tropical area, which lies between the tropics of Cancer and Capricorn, is characterized by mean annual rainfall above 250 mm and a mean temperature of 18°C in the coldest month (22°C following different authors in Thurston, 1984). Many patterns of rainfall distribution occur in this region, such as even distribution throughout the year (up to 2,000 mm/year), a dry and rainy season, or two wet and two dry periods. The length of the growing season (depending on the rainy season), its seasonal distribution and the amount of rainfall strongly influence the agricultural potential of tropical regions (Frere, 1984; Holm *et al.*, 1979).

Climatic diversity influences the distribution of cereal crops. The three main rainfed cereal crops found in tropical areas are millet, sorghum and maize (Table 8.1). The last two can also be grown under irrigation, like wheat which exists in some tropical highlands and sub-tropical regions.

Table 8.1. Production and yield of maize, sorghum and millet in developing countries in Africa, Latin America and the Far East. (FAO Production Yearbook, 1987.)

	Maize		*Sorghum*		*Millet*	
	area 10^3 *ha*	*yield kg/ha*	*area* 10^3 *ha*	*yield kg/ha*	*area* 10^3 *ha*	*yield kg/ha*
World	131475	3345	46807	1526	33977	626
Africa	14551	1300	11114	928	14886	711
Latin America	27190	1943	4366	2905	200	1200
Far East	2164	1531	16665	667	17093	534
sub-total area or mean yield for developing countries						
	43905	1710	32145	1054	32179	616

Maize

Zea mays L. is the second most important crop in the world in terms of total food production. It is the major cereal in Latin America (table 8.1). In tropical developing countries, it is used predominantly for direct human consumption. Its water requirements are high which limit its extension. Annual rainfall must be 600–1200 mm with adequate moisture during the rapid growth at midseason (Holm *et al.*, 1977). Irrigation is necessary in dry regions, hence maize is not often grown on small-scale farms in the semi-arid tropics where rainfall is low and erratic. In traditional farming systems, maize is rarely a monoculture as it is grown intercropped with pulses, squash or other cereals. Yield varies from less than 1000 kg/ha with low inputs, to 4000–5000 kg/ha with high inputs (Frere, 1984).

Sorghum

Sorghum bicolor (L.) Moench (= *S. vulgare* Pers.) is a major staple for people who live in the world's semi-arid tropics. In India and Africa it is grown traditionally, primarily as a grain, while in America it has been introduced quite recently as a livestock feed, produced with modern technology (Fernandes and Schaffert, 1978).

Because of its drought tolerance, sorghum may be found in regions where annual rainfall during the growing season is over 400 mm. Different varieties are adapted to the length of the growing season. Sorghum is grown on light sandy soils as well as on heavy clays. It is also widely grown along waterways in West Africa on receding moisture after floods (Sidibe, 1978). In Africa, sorghum is grown as a sole crop or intercropped with cowpea (*Vigna unguiculata* L.), groundnut (*Arachis hypogaea* L.) or millet.

In traditional farming, production with low inputs may be around 750–1000 kg/ha; with modern technology, it may reach more than 3000 kg (Mexico 3700 kg in 1986) and even more than 4000 kg with the intensive farming systems of Europe.

Millet

In the tropics, millets are fourth in importance among cereals after maize, rice and sorghum. The term 'millet' is used for a wide range of grasses from several genera with small edible seeds which do not shatter readily at maturity.

Major tropical species are pearl (or bulrush) millet, *Pennisetum typhoides* Stapf. = *P. glaucum* R.Br. = *P. americanum* (L.) Leeke, mainly cultivated in the semi-arid tropics in India and in Africa, and finger millet, *Eleusine coracana* (L.) Gaertn., grown in India and East Africa. Millets are able to grow in the dry rainfed areas (annual precipitation of 300–350 mm) on sandy, acidic and infertile soils (Harinarayana, 1987; Spencer and Sivakumar, 1987). Locally, and of less importance, foxtail millet or moha, *Setaria italica* (L.) P.B., can be found mostly in India; fonio, *Digitaria exilis* (Kippist.) Stapf., in West Africa and Teff, *Eragrostis tef* (Zucc.) Trotter in Ethiopia (Deuse and Lavabre, 1979; Holm *et al*. 1977; Mersie and Parker, 1983).

In temperate regions and occasionally in tropical areas (India), proso millet (*Panicum miliaceum* L.) and japanese millet (*Echinochloa crus-galli* var *frumentacea* (Roxb.) Wight), are planted as catch-crops (Cavers and Bough, 1985). Weed problems in pearl millet and finger millet have been quite well studied at ICRISAT (Shetty and Krantz, 1980) and will be considered later. Mean yield in the world is around 750 kg/ha; best production with high inputs may reach 1000–1500 kg/ha, though worst yields are between 500–600 kg, or even less (Niger: 425 kg/ha, Sudan: 366 kg/ha in 1986) (FAO, 1987).

Wheat

Among tropical countries, India, Pakistan, Brazil and Mexico are the major producers of wheat, mostly on irrigated lands in semi-arid or subtropical conditions and in the highlands (FAO, 1987). In Africa, 90% of the 8144×10^3 ha sown in 1986 is located in Mediterranean coast countries and Mauritania, Ethiopia and South Africa; Kenya and Sudan with 140×10^3 ha and 150×10^3 ha respectively produce small amounts of wheat. Night temperatures are often too high for adequate tillering in tropical regions and though production is expanding in countries like Senegal, Chad and Cameroon (Deuse and Lavabre, 1979), wheat can not really be considered as a 'tropical cereal'.

The dwarf wheat varieties with high yield potential which have been developed in recent years (Tandon, 1985) require high inputs. So in most of the countries, except in a few parts of India, wheat is mainly grown on estates or large farms. Unlike the previous cereal crops, seeds are broadcast, so it is difficult to eradicate weeds after sowing. For all these reasons, wheat is not a major small-farm crop and it will not be considerd in detail in this chapter.

Table 8.2. Most important weeds in tropical cereals (except wheat).

				Maize	*Sorghum*	*Millet*
PANICOIDEAE, Paniceae						
Brachiaria						
deflexa (Schumach.) Robyns	Af			*	*	
distachya (L.) Stapf	Af				*	
fasciculata (Sw.) Blake		Am		***	**	
lata (Schumach.) Hubb.	Af			*	*	
plantaginea (Link) Hitch.		Am		*	**	
ramosa (L.) Stapf	Af		As			
reptans (L.) Gard. and Hubb.		Am	As			
xantholeuca (Schinz) Stapf	Af					
Cenchrus						
biflorus Roxb.	Af	Am				*
brownii Roem. and Schult.		Am			*	
echinatus L.		Am	As	*		
pauciflorus Benth.	Af	Am				
Digitaria						
adscendens (H.B.K.) Henr.	Af		As			
bicornis (Lam.) London		Am				
ciliaris (Retz.) Koel.	Af	Am	As	***	**	*
fuscescens (Presl.) Henr.	Af			*		
horizontalis Willd.	Af	Am		**	*	
insularis (L.) Mez.ex Ekman.		Am				
sanguinalis (L.) Scop.	Af	Am	As	**	**	
velutina (Forssk.) P.Beauv.	Af					
Echinochloa						
colona (L.) Link	Af	Am	As	***	*	
crus-galli (L.) P.Beauv.		Am	As	*		
Ixophorus						
unisetus (Presl.) Schlech.		Am		*		
Panicum						
dichotomiflorum Michx.		Am	As	**	*	
fasciculatum Swartz.		Am	As			
hirticaule Presl.		Am		*		
laevifolium Hack.	Af					
laxum Sw.	Af					
maximum Jacq.	Af	Am	As			
repens L.	Af	Am	As			
trichoides Sw.		Am		*		
Paspalidium						
geminatum (Forssk.) Stapf	Af	Am	As			
Paspalum						
conjugatum Berg.		Am				
notatum Fluegge		Am				

Table 8.2. *Continued*

				Maize	Sorghum	Millet
orbiculare Forst.	Af			*		
paniculatum L.		Am		*		
paspaloides (Michx.) Scribn.		Am	As	*		
scrobiculatum L.	Af			*		
virgatum L.		Am				
Pennisetum				*		
clandestinum Chiov.	Af	Am	As		*	*
pedicellatum Trin.	Af		As		*	*
polystachion (L.) Schult.	Af		As		*	*
purpureum Schumach.	Af		As			
violaceum (Lam.) L. Rich.	Af					***
= *P. americanum* subsp. *monodii* (Maire) Brunken						
Rhynchelytrum						
repens (Willd.) Hubb.	Af	Am				
Setaria						
barbata (Lam.) Kunth		Am				
glauca (L.) P.Beauv.			As			
gracilis Kunth	Af	Am	As	*		
italica (L.) P.Beauv.	Af		As			
longiseta P.Beauv.	Af					
pallide-fusca Stapf. and Hubb.	Af			*	*	
palmifolia (Koen.) Stapf			As			
verticillata (L.) P.Beauv.	Af	Am	As	**	*	
Urochloa						
panicoides P.Beauv.	Af					
reptans (L.) Stapf		Am				
PANICOIDEAE Andropogoneae						
Andropogon						
bicornis L.		Am				
Heteropogon						
contortus (L.) Roem. and Schult.	Af			*		
Imperata						
cylindrica (L.) Raeuschel	Af	Am	As	**		
Rottboellia						
exaltata L.f.	Af	Am	As	**		
= *cochinchinensis* (Lour.) W.D. Clayton						
Schizachyrium						
exile (Hochst.) Pilg.	Af					
Sorghum						
bicolor (L.) Moench	Af		As			
halepense (L.) Pers.	Af	Am	As	**	**	
verticilliflorum (Steud.) Stapf.	Af	Am			**	

Table 8.2. *Continued*

				Maize	Sorghum	Millet
CHLORIDOIDEAE						
Aristida						
adscensionis L.	Af	Am				
ternipes Cav.		Am				
Chloris						
chloridae (Presl.) Hitch		Am		*		
gayana Kunth	Af	Am				
pilosa Schumach.	Af			*	*	**
prieurii Kunth	Af					
pycnothrix Trin.	Af					*
virgata Sw.	Af	Am		*		*
Cynodon						
dactylon (L.) Pers.	Af	Am	As	***	**	**
Dactyloctenium						
aegyptium (L.) Willd.	Af	Am	As	**	**	**
Eleusine						
indica (L.) Gaertn.	Af	Am	As	**	**	**
Eragrostis						
aspera (Jacq.) Nees	Af					*
cilianensis (All.) Lut.		Am				
ciliaris (L.) R.Br.	Af	Am			**	**
diplachnoides Steud.	Af					
elegantissima Chiov.	Af			*		*
gangetica (Roxb.) Stend.	Af					
pilosa (L.) P.Beauv.	Af	Am	As			
tenella (L.) Roem. and Schult.	Af					
tremula Steud.	Af		As			*
Leptochloa						
chinensis (L.) Nees		Am				
filiformis (Lam.) P.Beauv.		Am		*		
scabra Nees		Am				
virgata (L.) P.Beauv.		Am				
Sporobolus						
indicus (L.) R.Br.		Am				
jacquemontii Kunth	Af					

Footnotes:
Af = Africa, Am = America, As = Asia.
MA = maize, SO = sorghum, MI = millet.
* = locally abundant
** = frequent and abundant
*** = frequent and very abundant

Grassy weeds in tropical cereals

Main problematic species

The Poaceae family is represented in tropical areas predominantly by Paniceae and Chloridoideae (Häfliger and Scholtz, 1980). Pooideae, though mostly of temperate origin, have spread also in tropical areas where they are present essentially in the highlands (Popay and Ivens, 1982; Holm *et al*., 1977) or in irrigated wheat crops (Gill, 1986).

A list of the most important weeds of maize, sorghum and millet (Table 8.2) has been compiled from 180 species cited in the bibliography (see references quoted with *). Geographic indications do not mean that the species is present only in the mentioned continent or crop, but that it is described as a problematic weed in these conditions. Such a list can not be exhaustive and it reflects some subjective judgements. It is very difficult to find literature giving frequency and dominance of weeds in a region. Generally most authors present only predominant weeds. In Table 8.2 the 'world's worst tropical grasses' are written in bold whereas regionally harmful weedy grasses are underlined with a single line.

Sixty-two Panicoideae have been listed, of which 54 belong to 12 genera of Paniceae and eight to six genera of Andropogonae. Twenty six species of Chloridoideae belonging to eight genera are listed.

Some genera like *Digitaria*, *Echinochloa* or *Sorghum* are well represented in every continent, others are more important in America (e.g. *Paspalum* or *Leptochloa*), in Asia and Africa (e.g. *Pennisetum* and *Setaria*), in Africa and America (e.g. *Brachiaria* and *Eragrostis*), but many weeds have been introduced either deliberately or accidentally by man and have become established in every tropical region. Hashimoto (1982) explains that many Poaceae have been introduced into Brazil for the establishment of pastures and have gradually become weeds such as *Pennisetum clandestinum*,* now a weed of sorghum, or *Hyparrhenia rufa* (Nees) Stapf. found sometimes in maize of 'desmonte' (Leger, 1990). Popay and Ivens (1982) note that in East Africa and Zimbabwe, respectively, 20% and 32% of weeds are of American origin and were established in the early nineteenth century. The weeds associated with American crops (like maize and tobacco) presumably arrived in impure crop seeds. *Aristida adscensionis* and *Panicum laxum*, originating from America (Merlier and Montégut, 1982), are among major grass weeds in West Africa.

Introductions from Asia to Africa, and vice versa, are more difficult to explain, as exchanges between the two continents are very ancient. However some grassy weeds such as *Rottboellia cochinchinensis*, a native of Asia, have become widespread in the tropics quite recently (Mercado,

*Authors' names are indicated in Table 8.2 or mentioned in the text for other species.

1978). An exotic weed may reveal high noxious potential in its new habitat, such as *Sorghum halepense* or *Rottboellia cochinchinensis* in maize and sorghum of central America (Holm *et al*., 1977; Ambasht, 1982; Bridgemohan and Brathwaite, 1989).

The high phenotypic plasticity of many Chloridoideae, such as *Cynodon dactylon*, *Eleusine indica*, *Dactyloctenium aegyptium* or *Eragrostis ciliaris*, has allowed them to colonize most annual crops of tropical regions under a wide range of edapho-climatic conditions; they are now considered as cosmopolitan weeds (Krantz *et al*., 1977; Merlier and Montégut, 1982).

Crop–weedy grasses association

As stated by Popay and Ivens (1982), 'very little information is available on weeds in small-holder crops, including maize, sorghum and millet ... most publications refer only to a few species resistant to control measures'.

Most of the important grassy weeds of a region have detrimental effects in every crop. *Cynodon dactylon*, *Eleusine indica*, *Eragrostis ciliaris*, *Dactyloctenium aegyptium*, *Digitaria ciliaris* and *Cenchrus biflorus* are found in maize, sorghum and millet (Table 8.2). However some species have special ecological requirements, such as need of high moisture or good fertility, low tolerance of poorly drained soils, which limit their extension. For instance, some species, though geographically widespread, are restricted to maize fields because of favourable moisture conditions: *Rottboellia cochinchinensis*, *Imperata cylindrica* and *Panicum maximum* are rarely found in millet. By contrast, in India and Africa, many Chloridoideae are well adapted to sandy soils and are drought tolerant; for example *Chloris prieurii*, *C. pilosa*, *Eragrostis aspera* and *Aristida adscensionis*. These provide a large number of weeds of millet and do not occur readily in maize crops because they are not competitive with the crop or the other weeds. Marnotte (1984), in the Ivory Coast, distinguishes species by their ability to grow in plots with different fertilizer levels. *Eragrostis tremula* and *Digitaria fuscescens* are frequent on unfertilized plots while *Brachiaria lata* and *Eleusine indica* respond to high levels of fertilizer. In Senegal, Fontanel (1987a, b) confirms that *E. tremula* is typical of unfertile soils as are *E. cilianensis*, *Cenchrus biflorus*, *Brachiaria distichophylla* (Trin.) Stapf. and *Andropogon techorum* Schumach and Thonn.; whereas *Brachiaria lata* and *Eleusine indica* grow well in fertilized fields and are favoured by increased manure, as are *Setaria verticillata*, *Pennisetum pedicellatum*, *Digitaria ciliaris*, *Chloris pilosa* and *Dactyloctenium aegyptium*.

Grassy weeds are a significant component of tropical weedy communities. Jurgens and O'Reilly (1975) note that grasses are the most important of species in maize fields of the Dominican Republic. Beck

(1985), in a survey of weedy species in maize in Nicaragua, found they totalled 24.5% of 179 species; grasses represented respectively 15.6 and 16.5% of 200 weeds in maize intercropped with lemons (Aguilar *et al.*, 1986) or alone (Maillet, 1991). In Trinidad, the composition of the weed flora in unweeded plots of maize averaged 11–12%, 69–74% and 10–15% for broadleaves, grasses and sedges, respectively (Bridgemohan and Brathwaite, 1989). In these surveys, grasses were also the most abundant in density and biomass. In Africa, Traore (1987) gives percentages higher than 15% for grassy weeds of maize in Ivory Coast. Many authors (Deuse, 1986; Ayeni *et al.*, 1984; Marnotte, 1984; Popay and Ivens, 1982) emphasize the role of Poaceae in cereal agro-ecosystems, especially in the semi-arid tropics.

Associated with wheat, a few temperate Pooideae may be found mixed with local weeds: *Alopecurus myosuroides* Huds., *Avena fatua* L., *A. sterilis* L., *Bromus diandrus* Roth., *B. secalinus* L., *Lolium rigidum* Gaud., *L. temulentum* L., *Phalaris minor* Retz and *Polypogon monspeliensis* (L.) Desf. are important weeds, whether in India, Kenya, Uganda or Colombia (Ambasht, 1982; Popay and Ivens, 1982; Häfliger and Scholtz, 1980). *Setaria verticillata*, *S. pallide-fusca*, *Cynodon dactylon*, *Digitaria scalarum* (Schweinf.) Chiov. and *Eleusine indica* have also been able to colonize wheat fields and to compete with European species (Bogdan, 1965). In Ethiopia *Snowdenia polystachya* (Fres.) Pilg, a short-lived perennial, has become an important weed (Parker and Fryer, 1975; Ivens, 1989).

Mimic weeds

Grass weeds are often difficult to eradicate in cereal crops because they have many similarities with the crop. They generally have the same life cycle and they may also mimic the crop vegetatively, so they are not easily identified until maturity and are difficult to eliminate by hand-weeding.

Ivens (1989) emphasizes the difficulties of controlling *Eleusine indica* subsp. *indica* and *E. indica* (L.) Gaertn. subsp. *africana* Phillips, a robust tetraploid plant, in finger millet fields. According to Reid and Silawve (1987), the weedy species are easily recognized by small farmers at the flowering stage, but not at the vegetative stage. In India, more than 50% of *Phalaris minor* seedlings escape hand-removal or hand-hoeing in wheat fields because of their morphological similarity to the crop (Gill, 1986). Furthermore, if weeds are related genetically to the cultivated races they are likely to hybridize with them.

In Ethiopia and Sudan (Harlan, 1982), Dominican Republic (Jurgens and O'Reilly, 1975), and Latin America in general (Häfliger and Scholtz, 1980), the wild plant *Sorghum verticilliflorum* crosses readily with cultivated varieties of sorghum; the hybrids are fertile and produce spikelets which shatter at maturity. They have become a serious pest in the USA as has 'shattercane', a race derived directly from cultivated sorghum.

Pearl millet has a high percentage of cross-pollination. In its centre of diversity and origin (Sahelian Africa), it freely crosses with its wild progenitors *Pennisetum violaceum* and *P. fallax*, producing fertile hybrids called 'shibra', *P. americanum* subsp. *stenostachyum* (Kletzsh ex A.Br.) Brunken (Williams and Andrew, 1983; Spencer and Sivakumar, 1987). It also crosses with *Cenchrus ciliaris* L. and more distantly related species within *Pennisetum* (*P. pedicellatum*, *P. polystachyon*). These mimetic weeds escape removal during weeding and persist in the crop 'allowing for extensive modification of the genome of the cereal' (de Wet, 1987). At maturity, shibra heads shatter and disperse their seeds to reinfest the field. William and Andrew estimate that shibra may represent up to 20% of plant populations in West African pearl millet fields. Grazing may favour the wild species, as shown by Aken'ova (1985) in Nigeria.

Introgression occurs also between foxtail millet and weedy *Setaria* whenever the crop is grown in the proximity of its wild relatives (Darmency *et al*., 1984).

Weeds and crops may have similar physiological properties, such as mechanisms of herbicide detoxification. A large number of Panicoideae are able to partially detoxify atrazine by non-enzymatic hydrolysis to hydroxyatrazine and by enzymatic conjugation of atrazine to glutathione (Bhowid and Nandihally, 1988). In large-scale farming or on experimental stations increased abundance of *Digitaria*, *Panicum*, *Brachiaria*, *Setaria* and *Rottboellia* species in maize and sorghum fields is induced by continuous use of triazine herbicides (Deuse and Lavabre, 1979; Marnotte and Tehia, 1983; Bridgemohan and Brathwaite, 1989).

Biology of weedy grasses

Studies related to weed biology are few and incomplete. Basic research is necessary to understand better the mechanisms that determine weed proliferation and to choose a judicious weeding method according to each situation. Identification of exploitable biological differences between the crops and the weeds will allow an enhanced management programme (Andres and Clements, 1984).

Seed production and longevity in annual grasses

According to Kellman (1974), seed bank densities and dominance by weedy species (especially grasses) increase rapidly with continuous farming in tropical conditions. Some observations on seed production confirm that weedy grasses have the ability to create huge seed banks in the soil (Table 8.3): these represent the most important way of re-infestation. Table 8.4 (from Garwood, 1989) presents results of studies done in tropical arable

Table 8.3. Seed production of some weedy grasses.

Species	*Seed production*		*Author*
	/m²	*/plant*	
Eleusine indica	422,000–543,000	400–4,000	Schwerzel 1970
Rottboellia cochinchinensis	13,000–28,000	2,200	Schwerzel 1970
Dactyloctenium aegyptium		66,000	Holm *et al*. 1977
Echinochloa colona		3,000–6,000	Holm *et al*. 1977

Table 8.4. Seed bank density in tropical arable land.

Region		*Seeds/m²*
Florida	farm field	4,250
Belize	maize field	5,200–6,760
Venezuela	field after slashing	126
Senegal	crop field	5,029–7,683
Ghana	crop field	4,900
Nigeria	crop field	2,240
Kenya	crop field	2,290

Table 8.5. Potential viability of weedy grasses from cereal crops.

Species	*Longevity*	*Author*	
Eleusine indica	5–8 years	Hawton and Drennan	1980
Echinochloa crus-galli	13 years	Dawson and Burns	1975
Setaria lutescens	13 years	Dawson and Burns	1975
Setaria viridis	13 years	Dawson and Burns	1975
Digitaria ciliaris	1.5 years	Marks and Nwachuku	1986
Eragrostis gangetica	1.0 year	Marks and Nwachuku	1986
Paspalum orbiculare	1.0 year	Marks and Nwachuku	1986
Imperata cylindrica	1.0 year	Brook	1989
Rottboellia cochinchinensis	3.0 years	Mercado	1978
Cynodon dactylon	2.0 years	Rochecouste	1962

land. Though not specific to cereal crops, except in Belize, these estimations give an idea of mean seed density in the soil. Under monoculture in temperate regions, where seed longevity is much greater, weed densities may reach 7,000–30,000/m^2 (Lopez *et al.*, 1988; Schweizer *et al.*, 1989).

The seed bank is a function of the longevity of seeds in soil and is determined by the inherent viability of seeds and dormancy mechanisms. Data on longevity of even the most pernicious weeds are scarce. An extensive survey of the germination of fresh seeds sown under suitable conditions shows that rapid germination is the commonest response in tropical conditions (Garwood, 1989). However, many weedy species may remain dormant for a few months and in some cases for a few years (Table 8.5). It appears that in most cases mortality is very high in the first six months (Marks and Nwachuku, 1986). Dawson and Burns (1975) observed a very high longevity for *Echinochloa* and *Setaria* species; but this occurred with populations from the USA; the same results may not be obtained with tropical populations.

Germination and growth requirements

The analysis of dormancy mechanisms and of seedling emergence is very important in the perspective of integrated weed management. It is a complex process which has not yet been elucidated and for which we possess fragmentary knowledge in tropical conditions.

Most of the seeds produced during the growing season or at the beginning of the dry season are dormant for a few weeks. A period of after-ripening may prevent the seeds from germinating in response to the occasional dry season showers. Dormancy may disappear after exposure to drought as shown by Fenner (1980) for *Aristida adscensionis*, *Chloris pycnothrix* and *Setaria verticillata*, or by Marks (1983) for *Paspalum orbiculare*, *Eragrostis gangetica* (Roxb.) Steud., *Digitaria ciliaris* and *D. longiflora* (Retz.) Pers. Germination is often promoted by light and high temperature, as noted for *Pennisetum pedicellatum* and *P. polystachyon* (Noda *et al*., 1985a) and the increased emergence of weed seedlings caused by soil disturbance is a well known phenomenon. Obscurity or leaf shade partially inhibits germination of numerous species (Marks and Nwachuku, 1986). In fact, there is always a high variability in dormancy even among one population; this is due to genetic as well as phenotypic diversity (Egley and Duke, 1985).

According to dormancy pattern and seed heterogeneity, grassy weeds may have either one flush of germination (homogeneous seed population with short innate dormancy and no induced dormancy) or successional germination throughout the growing season (heterogeneous population, with induced, cyclical dormancy or seeds of different ages). *Rottboellia cochinchinenesis*, *Digitaria ciliaris* and *Eleusine indica* have a peak of

Table 8.6. Effect of light reduction on dry weight of weeds in the Philippines.

Species	*Available light*	
	50%	20%
	Reduction in dry weight	
Rottboellia cochinchinensis	52%	93%
Digitaria sanguinalis	56%	95%
Eleusine indica	60%	90%

emergence early in the rainy season and small peaks later on, linked with rainfall (Zimdahl *et al*., 1988, Marks and Nwachuku, 1986).

In Senegal, Fontanel (1987a, b) distinguishes early germinating species in maize and millet. *Cenchrus biflorus*, *Chloris pilosa*, *C. prieurii* and *Setaria pallide-fusca* appear at the beginning of the rainy season, while *Pennisetum pedicellatum*, *Digitaria ciliaris*, *Dactyloctenium aegyptium*, *Eleusine indica* and *Brachiaria lata* have successive flushes of seedling emergence. Marnotte (1984) contrasts *Brachiaria lata*, which germinates during the whole rainy season in Ivory Coast with *Eragrostis* species (*E. ciliaris*, *E. tremula*, *E. aspera*) which appear at the end of the growing season and thus are not very troublesome in maize.

Weed life cycles may be extremely short in tropical areas. Plants such as *Echinochloa colona*, *Setaria verticillata* or *Dactyloctenium aegyptium* may achieve production of mature seeds within 30–45 days under short-day photoperiods. Rapid seed set is not an absolute requirement for success as a weed: *Rottboellia cochinchinensis* and *Eleusine indica* need two to four months to produce seeds (Ambasht, 1982; Holm *et al*., 1977; Popay and Ivens, 1982). *Rottboellia cochinchinensis* that germinates with maize in the Philippines and escapes weeding can begin to shed within seven weeks and continue until maize harvest at 14 weeks (Fisher *et al*, 1985).

Most tropical weedy Paniceae and Chloridoideae are C4 plants; the C4 pathway appears to be of adaptative value in environments characterized by high light, high temperature and moderate moisture (Patterson, 1985). Shading effects may effectively reduce their biomass production as shown by Bantilan *et al*. in William and Chiang (1980)(Table 8.6), though species such as *Rottboellia cochinchinensis* maintain the capacity for rapid growth when later exposed to intense solar radiation (Fisher *et al*., 1985).

Perennial grasses

Some of the most troublesome weeds of tropical cereal crops are rhizomatous or stoloniferous weeds (Holm *et al*., 1977). Vegetative reproduction allows rapid growth and early colonization in the field with high competitive effects.

In small farms perennial grasses used not to be predominant because of intensive hand-weeding, low fertilization and long fallow periods prejudicial to species such as *Cynodon dactylon* and *Imperata cylindrica* (Parker and Fryer, 1975). With an increase in sedentary farming and greater levels of fertilizer, these species and others, like *Sorghum halepense* or *Setaria gracilis*, are now among the worst weeds of maize and sorghum.

Fragmentation and burial of subterranean organs of perennial grasses is a consequence of cultivation. The more intensive the cultivation, the greater is the frequency of smaller fragments (Kiger and Keller, 1985). The probability of successful sprouting and establishment decreases with depth of burial but increases with size of the segment as shown for *Cynodon dactylon* or *Panicum repens* (Moreira, 1978), *Sorghum halepense* (McWhorter, 1972), and for *Imperata cylindrica* (Brook, 1989). However, buried rhizomes of *Cynodon dactylon* (Fleury, 1969) may sprout readily from 30 cm depth and those of *Imperata cylindrica* from 15 cm (Brook, 1989).

The vegetative propagules of most weeds are less resistant to desiccation than seeds; rhizome segments of *Cynodon dactylon* dry rapidly if exposed to the air and viability is lost when water content is lower than 15–20% (Thomas, 1969). As *Cynodon* cannot survive long periods of dryness, it is only really a problem in irrigated areas and the humid tropics (Krantz *et al*., 1977). In *Cynodon dactylon*, as in *Pennisetum clandestinum*, an antagonism exists between vegetative and sexual reproduction (Fleury, 1969); however, perennial grasses generally exploit both possibilities in order to reproduce and spread. Vegetative reproduction is often affected by a decrease in the daily input of light energy (Patterson, 1980). The relative allocation of biomass to rhizome growth under weak light intensity is reduced in *Imperata cylindrica* and *Sorghum halepense* (Patterson, 1980; McWhorter and Jordan, 1976).

Crop losses due to grassy weeds

Direct losses

It is quite difficult to separate losses due to grasses, sedges or broadleaved weeds. In the literature, weed competition is generally estimated with reference to the whole community. Some information exists, however, on the effects of some weedy grasses in maize (Table 8.7). It is easier to determine the critical period of competition for each crop. Generally crops 'seldom suffer from weed competition during the very early phases of development, and weeds emerging after a certain stage of crop development no longer affect yields' (Koch *et al*., 1982).

Table 8.7. Yield losses due to grassy weeds in maize.

Species	*Density*	*Loss*	*Author*	
Panicum dichotomiflorum	12–20 plants/m^2	15–77%	FAO	1982
Setaria faberi	13.1 plants/m of row	18%	Beckett *et al.*	1988
Rottboellia cochinchinensis	13 plants/m^2	33%	Popay and Ivens	1982
	50 plants/m^2	47%	Popay and Ivens	1982
	142 plants/m^2	71%	Popay and Ivens	1982

The weed-free period recommended for maize in tropical countries varies a little with the type of flora, the productivity of the site and the density of the crop. Most authors say that weeds should be controlled from the 10th to 30–40th day after sowing in the case of annual weed communities and from the first week where perennial weeds are present (Nieto *et al.*, 1968; Schmidt *et al.*, 1983; Zimdahl, 1980).

Shen and Peters compared the effect of in-row and inter-row competition in the USA; they found that maize yields were more reduced by in-row competition (FAO, 1982) which is corroborated by most field observations (Traore, 1987; Marnotte, 1984; Akobundu, 1982; Fisher *et al.*, 1985).

Sorghum has the same requirements as maize. In Sudan, losses of 40–80% have been observed (Koch *et al.*, 1982; Deat, 1986). In Colombia, fields unweeded during the first 20 days but weeded thereafter were able to recover from the effects of competition; this was impossible with competition throughout the first 30–40 days (FAO, 1982).

Harinayarana (1987) estimates that in India, competition from weeds could reduce pearl millet yield by 25–50%. The critical period is 20–40 days after sowing according to Balyan *et al.* (1985). In finger millet, weeding should not be delayed beyond five weeks after sowing and a clean field should be maintained thereafter. In unweeded plots, yield losses reach 40% and grasses are more competitive than broadleaved weeds (Hedge *et al.*, 1983). *Phalaris minor* may decrease wheat yields by 36–80% in unweeded fields in India (Gill, 1986). Effects of competition have already been described in Zimdahl (1980) and in a review of weed interference in tropical cereal crops (FAO 1982).

Grassy weeds also play another damaging role in that they are often secondary plant hosts for insects or pathogens. Thus larvae of *Spodoptera* spp. are likely to develop on wild grasses in the field and the millet midge (*Geromya penniseti*) also lives on *Pennisetum pedicellatum*, *Setaria pallide-fusca* and *Echinochloa colona* (Mbaye Ndoye and Gahukar, 1987). *Setaria verticillata* serves as an alternative host for downy mildew of pearl millet (*Scleropora graminicola* Sacc.) and ergot has been observed on *Cenchrus ciliaris* and *Panicum antidotale* Retz. (Harinayarana, 1987).

Sorghum halepense is an alternative host of sorghum midge in the USA and Mexico (Ramaiah and Parker, 1982). In the Philippines, downy mildew of maize (*Peronosclerospora philippinensis*) occurs on *Saccharum spontaneum* L. (wild sugarcane).

Indirect losses

Besides reducing the yield, weedy grasses also cause problems by injuring the farmer (spiny spikelets of *Cenchrus echinatus* or brittle hairs on the sheath of *Rottboellia cochinchinensis*) and contaminating the grain, thereby reducing its value in the market.

Parker and Fryer (1975) emphasize the limitation of cultivated area and sowing due to land preparation and weeding. In Zambia, hand-weeding of finger millet takes 32 days/ha (Reid and Silawve, 1987); in Cameroon, manual weeding takes at least 15 days/ha in maize, millet or sorghum (Chabrier, 1986). Gill (1986) reports that hand-hoeing in wheat fields of India takes 10–16 days/ha.

Although weeds are sometimes advantageous in agro-ecosystems (Swamy and Ramakrishnan, 1988; Maillet, 1991; Chacon and Gliessman, 1982), their detrimental effects are generally greatest in small farmer subsistence systems.

Weed management in tropical cereals

From the above, it is obvious that grassy weeds are characterized by a huge diversity. They have so many highly divergent life cycles, varying periods of germination, different ecophysiological requirements and distinct patterns of biogeographical distribution, that weed management obviously has to be decided case by case: a single method of weed control is inappropriate.

However, some particular points stand out which provide guidelines for weed management:

- results of most experiments emphasize the importance of early weeding which corresponds with peaks of germination of many weeds and critical periods of competition for the crop; this period is also the busiest one and the weeding is one of the critical labour bottlenecks for small farmers;
- a second weeding is needed before maturation of weed seeds in order to prevent dissemination;
- weeding within the row is a priority;
- decrease of fallow periods and a repeated monoculture induces increasing density of weeds;
- changes in technical practices may result in weed shifts;

- invasion processes between continents continue to play a determinant role.

As pointed out by Shaw (1982), weed management has also to be considered on the basis of its 'economic, ecological and sociological consequences'. This means that the choice of weed control methods not only depends on technical solutions but relies also on other criteria. Advantages or disadvantages of specific weed control practices may be very different from one place to another relative to established systems; if integrated weed management programmes are technologically sound but uneconomical, farmers will not accept them. For all these reasons, integrated weed management cannot be discussed in isolation from other factors.

As there is a close relationship between weeds and crops, 'modern weed management strategies involve combinations of crop production practices and specific weed control technologies intended to reduce weed competition, thereby shifting the competitive balance in favour of the crop' (William and Chiang, 1980).

Cultural improvements

Any practice which benefits the crop by enhancing its competitive ability against weeds has to be considered. Seedling vigour, growth rate, early rooting, tillering habit and plant height give competitive advantages to the crop (Wall, 1986; Zimdahl, 1980). Early maturing varieties may avoid competition in comparison with those that are late maturing which are at a vulnerable period in their growth cycle when subjected to competition (Walker and Buchanan 1982). Selection of local varieties or hybrids of sorghum and pearl millet has been done by ICRISAT (Harinarayana, 1987).

Crop density may also help to reduce weed growth. In India, the optimum plant population is 180,000 plants/ha for sorghum and pearl millet (Shetty and Krantz, 1980) while in Zambia, Boringher (1987) recommends 100,000 plants/ha to obtain top yield. Farmers barely maintain 50–75% of the recommended density. Under high moisture conditions, grain sorghum competes best with weeds when planted densely in narrow rows; with less moisture, seeding in wider rows results in more effective competition (Wiese *et al.*, in William and Chiang, 1980). The influence of row spacing, seeding rate and canopy on crop–weed relationship has not been investigated enough; judicious manipulations of these factors could be a highly effective component of integrated weed management.

Improved varieties often need higher rates of fertilizer than usual to express their potential. African soils generally have a low phosphorous content and nitrogen management can be difficult because the soil has a low buffering capacity (Spencer and Sivakumar, 1987). The practice of recycling crop residues by incorporating them as a compost helps to replenish P and N levels, and maintain yield in the absence of fallow periods (Mattlon, 1987). Increase of manure or of chemical fertilizers must be linked to efficient weeding, for weeds may benefit more from fertilizer than the crop, thus reducing final yield (FAO, 1982; Fontanel, 1987a).

Residue incorporation, or mulch with cereal straw may be of interest because of the allelopathic influence it may have. According to Purvis *et al.* (1985), *Sorghum bicolor* residues exhibit selective effects on weed germination or growth under field conditions. The concept of selective cultivars with allelopathic properties is advanced by different authors (Putnam and Duke, 1978; Einhalling and Leather, 1988). Fontanel (1987a) reports that a previous year's crop of sorghum inhibits *Pennisetum pedicellatum* and *Dactyloctenium aegyptium*, but enhances *Eleusine indica* and *Digitaria exilis* in the next crop or fallow.

Monocropping and repeated monoculture often cause a rapid weed shift towards highly competitive species. In Trinidad or in the Philippines, maize may be planted 2–3 times a year; *Rottboellia cochinchinensis*, because of its competitiveness and similarity to maize in growth requirements, becomes rapidly dominant (Bridgemohan and Brathwaite, 1989; Fisher *et al.*, 1985). Forcella and Lindstrom (1988) found that in continuous maize cultivation with conventional tillage, the seed bank was around 500–1000 seeds/m^2 (with grassy weeds predominant), while in a maize–soyabean rotation, the soil seed bank was 200–700 seeds/m^2.

Hence a judicious crop rotation can achieve some degree of weed control. A spreading cultivar of pigeon pea (*Cajanus cajan* L.) in rotation with sorghum can effectively reduce shade-intolerant weeds such as many C4 grassy weeds (Shetty, 1976 in William and Chiang, 1980). Rotation of maize with crops in which aniline herbicides or specific antigraminaceous herbicides are used (groundnuts, cotton, soyabeans) has been suggested to chemically eliminate grassy weed species (Deat, 1986; Grichar and Bostwell, 1985).

If a long fallow period induces a real decrease of weed inoculum, a short fallow is not sufficient to deplete a soil seed bank (Marks, 1983) and can even encourage the accumulation of weed seeds from *Chloris pilosa*, *Cenchrus biflorus* or *Pennisetum pedicellatum* in the Sahelian tropics (Fontanel, 1987b).

Control by hand-weeding

Hand labour is very suitable as young seedlings emerge, but it is time-consuming so in many countries it is only done to the extent that family labour can permit and is rarely undertaken when most effective. It is still the commonest method though it becomes cost-prohibitive in many areas because of the gradually shrinking labour pool (Akobundu, 1980;

Montillet, 1988). Hand-weeding is useful in rows where mechanical weeding is not effective; the thinning of millet or sorghum in African semi-arid areas is done in this way (Mattlon 1987). Some weeds, such as *Panicum maximum* or *Eleusine indica*, that are chopped or pulled out may re-establish themselves quickly in wet conditions if not completely extracted (Ayeni *et al.*, 1984), and stoloniferous or rhizomatous weeds such as *Paspalum conjugatum*, *P. paniculatum*, *Sorghum halepense* and *Cynodon dactylon* (Beck, 1985; Rao *et al.*, 1985) resist hand-hoeing. In the absence of perennial weeds in the row, a good yield can be obtained if weeding is timely and frequently enough (Matthews, 1984). Control of shibra millet needs a combination of cultural and plant breeding activities. Shibra plants should be cut prior to anther shedding, and shibra gene contamination needs to be eliminated from local cultivars through breeding (Fussell *et al.*, 1987).

Mechanical cultivation

In small farming systems, mechanical cultivation is usually done by animal traction, rarely by tractor. Deep ploughing by tractor (at 20 cm) does not seem to be more effective on grassy weeds than shallow ploughing with cow-traction in the light soil of Senegal (Fontanel, 1987b). The use of animals decreases from the sub-humid to the semi-arid zones (Spencer and Sivakumar, 1987); in the humid tropics, soils are generally too wet during the rainy season to permit thorough inter-row weeding.

Many disadvantages of mechanical cultivation have been pointed out. Akobundu (1982) emphasizes increased risks of erosion due to heavy rains. Most authors report that ploughing is not effective to eradicate weeds between crop plants in the row and may damage crop roots, while weeds in drilled and broadcast crops cannot be controlled adequately without injuring the crop. Another disadvantage may be that buried dormant seeds are brought back to the surface and can give new generations of seedlings, though, according to Matthews (1984), weed removal in the inter-row is not of prime importance.

A combination of specific cultural practices, mechanical cultivation and timely hand-hoeing has proved to be very effective as demonstrated by Traore (1987). On flat land, the percentage of weed recovery reached 60% 25 days after sowing, while on simple and double ridge plantings, it reached 23% and 40% respectively. Optimum control of weeds is obtained with simple ridging that buries the superficial seeds associated with a first hand-weeding in the row 10 days after sowing and with a second weeding at 40 days.

Generally, the succession of two inter-row weedings at 2–3 weeks and then 5–7 weeks, according to weed growth, combined with one or two hand-weedings in the row gives good results and increases the yield of any cereal crop (Brook, 1989; Bridgemohan and Brathwaite, 1985; FAO, 1982).

Chemical weeding

Pre-emergence herbicides are now available to control grassy weeds in any tropical cereal (Deuse and Lavabre, 1979; Chabrier, 1986; Marnotte and Tehia, 1989; Bridgemohan and Brathwaite, 1980; Fisher *et al.*, 1985; Akobundu, 1982). Their use would be appropriate not only for minimizing early weed competition, but also for reducing the work load during the peak labour period; they avoid at least one or two cultivations during the first 3–4 weeks and control weeds in the inter-row as well as within the row. The use of herbicides on heavy clay soils makes it possible to overcome the problems of ploughing in the wet season (Seth, 1986).

Herbicides allow no-tillage practices with the additional use of a pre-plant herbicide such as paraquat to desiccate previous vegetation before sowing; no-till farming saves time and prevents soil erosion at planting (Akobundu, 1980). An integrated approach involving the use of herbicides and organic mulches may contribute to suppress germination of many weedy grass seeds and limit the growth of C4 shade intolerant weeds. Straw mulch must be thick to be effective against harmful weeds such as *Rottboellia cochinchinensis* (Bridgemohan and Brathwaite, 1989). However, perennial weeds, including grassy weeds, survive paraquat applications and efficient systemic herbicides such as glyphosate can be costly at the small-farm level.

The weed shift toward C4 grasses induced some years ago by the continuous use of triazine herbicides, may be stopped by combining herbicides such as alachlor, metachlor, bentazone and trifluralin with atrazine or simazine, terbuthylazine or cyanazine. Herbicide combinations are effective against a broader spectrum of weeds and often require the use of lower rates than when the chemicals are used singly. A rotation of herbicides should avoid the appearance of resistance (Deat, 1986).

Finally, the Ultra Low Volume spraying option provides the opportunity to spray only 20–40 l/ha (no problem of water accessibility) with light equipment. Absolute reliance on chemical weed control does not, however, seem possible for small-scale farming in the tropics; even when technically suitable, it may not be a practicable solution because of educational or economic limitations, or lack of a distribution system.

Where labour is abundant and small farmers have limited capital and inadequate training, manual methods are more economical. If the cost of hand-weeding is high, herbicide treatment becomes a competitive and promising way to control weeds, at least in the first few weeks. For instance, in India in 1970, the cost of an atrazine treatment was only half that of hand-weeding plus two hoeings (FAO, 1982). Although very effective against *Rottboellia* in maize in Trinidad, pendimethalin is too costly for small farmers (Bridgemohan and Brathwaite, 1989). Similarly, total production costs were greater than net returns with an intensive herbicide treatment against *Rottboellia* in the Philippines in comparison with traditional hand-weeding (Fisher *et al*., 1985). The highest net returns in cultivation of finger millet is given by two hand-weedings followed by 0.5 kg/ha nitrofen + 2 kg/ha propanil in an experiment in India (Boopathi *et al.*, 1985).

Fontanel (1987a), though convinced that herbicides would save time, recognizes that the costs of treatment and equipment are too high to be considered in Senegal. Low cost practices provided high net returns, whereas expensive treatments produce low net returns.

Biological control

This method has been successfully used against naturalized weeds in Australia and the USA. Its development could be of interest to deplete weed populations which are not effectively controlled by conventional means. As biological control focuses on a sole target species, this should be a dominant weed of the community. The great richness of tropical weed communities is not well adapted to biological control, except in a few cases of monospecific invasion as with *Imperata cylindrica* or *Rottboellia cochinchinensis.* Evans (1987) presents several 'candidate biological control agents' for both species, but further information is needed. Mercado (1978) mentions however that the helmeted guinea fowl (*Numida meleagris*) consumes large amounts of the seeds of *Rottboellia* in Zimbabwe. The only case of effective biological control of a grassy weed is the use of *Cochliobolus lunatus* Nelson and Haasis, an ascomycete, to control *Echinochloa crus-galli* in the Netherlands (Julien, 1987). More research is needed on biotic agents against perennial weeds which are especially difficult to eradicate without expensive systemic herbicides: target weeds could be *Cynodon dactylon, Sorghum halepense* or *Pennisetum pedicellatum*.

Multi-cropping (or intercropping) systems

Multi-cropping occurs widely in tropical areas and is practised essentially by small farmers (Ayeni *et al*., 1984). Studies on mixed cropping systems reveal that many biological factors, such as crop species, varieties, plant population and row arrangement, influence the nature and extent of weed growth. Most intercropping systems are designed to utilize the spatial arrangement and time dimensions more completely.

Increased cash value and highest land equivalent ratio occur generally if the association is well chosen. For instance, in India, pearl millet associated with mungbean (*Vigna radiata* L.) is economically advantageous (Harinayarana, 1987); the introduction in sorghum of early maturing crops such as cowpea (*Vigna unguiculata* L.), not only reduces weeding costs but also provides substantial additional income. Pigeon pea (*Cajanus cajan* L.) or mungbean associated with maize or sorghum suppress weed growth more effectively than the crop does when planted alone or intercropped with another cereal. The production of a rapid leaf area index inhibits the growth of Panicoideae and Chloridoideae: the potential biomass of *Eleusine indica* and *Digitaria sanguinalis* in maize alone was reduced by 66 and 50% respectively in maize intercropped with mungbean in the Philippines (William and Chiang, 1980).

Very often crop densities are too low and too much space is left at the beginning of the season. Unamma *et al*. (1986) found in Nigeria for the common cassava/maize association, that the highest economic returns were obtained from using cultural control involving high planting densities of low growing cowpea or egusi (*Colocynthis citrullus* (L.) O. Ktze); 'these two measures gave 78% and 84% greater economic returns respectively compared with the current recommendation of two hand-hoeings'. But Ayeni *et al.* (1984) emphasize the necessity to chemically control weedy grasses in a maize/cowpea association under low moisture in the early season.

The availability of new herbicides is thus interesting to help the establishment of the mixed crop; pre-emergence herbicides have been used successfully to enhance weed control in maize/chickpea (*Cicer arietinum* L.) and in sorghum/pigeon pea associations (Rao *et al*., 1985); the better development of both crops allows them to dominate *Cynodon dactylon*. Chloramben or alachlor controlled more than 80% of weeds in Unamma's experiments (1986), but generally these products are too expensive for small farmers. Bridgemohan and Brathwaite (1989) estimate that, in Trinidad, mungbean used as a living mulch to control *Rottboellia*, is only suited to the situation where small-scale farmers grow maize at wide spacings and low plant densities.

Though multi-cropping appears in most cases to decrease growth of weedy grasses and weeding time, it has to be combined with hand-hoeing or with chemical treatment to be really effective.

Conclusions

The evolution towards more settled agriculture and the need for increased food production requires technical improvements at the small-farmer level.

Though grassy weeds are dominant in the weed flora of tropical cereals, it is evident that integrated weed management should not be limited to grasses alone. Any programme of weed control must consider the whole weed community, including sedges and broadleaved plants as well as grasses. Except for chemical and biological weeding, which may be more specific but is not necessarily best adapted to subsistence farming systems, most control methods are aimed at controlling all weeds.

Even within the grasses, there is such a great diversity that integrated weed management has to be considered case by case. The intrinsic dynamics of a weed community induce relative shifts within the hierarchy of dominance among species; an increase of nitrogen, the depth of ploughing, the use of one type of herbicide or the kind of crop rotation may influence abundance and productivity of grassy weeds in different ways. The high phenotypic or genotypic diversity of species such as *Rottboellia cochinchinensis* or *Cynodon dactylon* allow selection at the level of the population too. Hence a weed management programme should integrate these data, at least when they are known, which is not very frequent.

The control of weeds must be done as early as possible, and particularly within the row; chemical weeding is well adapted to local treatments even with unsophisticated spraying material and should be promoted. Repeated hand-weeding is highly time-consuming and though effective if well done, constitutes probably the first hurdle to production.

The high cost of herbicide treatments is certainly a major obstacle, but some developing countries have created special facilities to enhance their use. The errors committed in developed countries should be avoided by an assessment of environmental risks linked to herbicide use in tropical conditions.

The biology and ecology of tropical weeds and particularly grassy weeds, is still not well known except for a few dominant species. Studies on weed seed banks would be very interesting as seed longevity is quite short in comparison with temperate species. Minimum tillage, which is interesting for many reasons, together with mulch, mixed cropping systems and well-timed hand-hoeing, might allow rapid depletion of seed banks.

Shading prevents germination as well as growth of C4 grassy weeds; every method or cultural practice which favours crop growth and enhances the establishment of a dense canopy is of interest. Mulches and multi-cropping systems should be completely integrated into weed management programmes.

Perennial grassy weeds often escape any kind of weeding; repeated hand-mowing can deplete their reserves during the growing season, but cultivation during the dry season might facilitate desiccation of fragments. Specific herbicides are too costly for small farmers, even in cash-crops. Biological control might be one solution.

Integrated weed management, defined as the combination of 'two or more weed control methods at low input levels in order to reduce weed competition in a given cropping system below an economical threshold level' (Akobundu, 1980) has proved to be a valuable concept in a few cases; much has still to be done to extend it to the small-farmer level.

References

* See page 120.

*Aguilar, J.V., Maillet, J. and Ramos, M.A. (1986) Estudio de las malezas en el agroecosistema maiz de riego asociado con frutales en Pueblo-Juarez. *Resumenes del V congresso de la ciencia en malezas,* Mexico, UAAN, 2pp.

Aken'ova, M.E. (1985) Effects of grazing on persistence and genotypic composition of an indigenous Nigerian pearl millet (*Pennisetum americanum* (L.) Leeke) population. *Weed Abstracts 1986* 35 (7), ref 2490.

Akobundu, I.O. (1980) Weed science research at the International Institute of Tropical Agriculture and research needs in Africa. *Weed Science* 28 (4), 439–45.

Akobundu, I.O. (1982) Rôle des techniques culturales de conservation des sols dans le cadre de la lutte raisonnée contre les adventices. *Bulletin Phytosanitaire FAO* 44, 29–52.

*Ambasht, R.S. (1982) The agrestal weed flora of India. In: Holzner, W. and Numata, N. (eds) *Biology and Ecology of Weeds*. Dr. W. Junk Publishers, The Hague, pp 267–76.

Andres, L.A. and Clements S.L. (1984) Opportunities for reducing chemical inputs for weed control. In: ASA, CSSA, SSSA (eds) *Organic Farming: Current Technology and its Role in a Sustainable Agriculture.* Madison pp 129–40.

*Ayeni, A.O., Duke, W.B. and Akobundu, I.O. (1984) Weed interference in maize, cowpea and cowpea/maize intercrop in a subhumid tropical environment. I. Influence of cropping system. *Weed Research* 24, 269–79.

Balyan, R.S., Bhan, V.M. and Malik, R.K. (1985) Effects of time of weed removal on growth and yield of pearl millet. *Weed Abstracts* 1986, 35 (4), ref. 1272.

*Beck, J. (1985) Die segetalflora im maisenbau nicaragua; untersuchungen über vorkommen, bekämpfung und interferenz mit der kulturpflanze. *Plits* 3 (4), Universität Hohenheim, Stuttgart, 88pp.

Beckett, T.H., Stoller, E.W. and Wax, L.M. (1988) Interference of four annual weeds in corn (*Zea mays*). *Weed Science* 36 (6), 764–9.

Bhowmik, P.C. and Nandihalli, B. (1988). Control of large crabgrass (*Digitaria sanguinalis*) and fall Panicum (*P. dichotomiflorum*) in corn with tridiphane and atrazine combinations. *Weed Science* 36 (3), 359–62.

*Bogdan, A.V. (1965) Weeds in Kenya wheat. *Weed Research* 5, 351–2.

Boopathi, S.N., Kolandaisamy, S. and Panchanathan, R.M. (1985) Economics of different weed control methods in finger millet. *Weed Abstracts* 1986, 35 (5), ref. 1792.

Boringher, A. (1987) Research on sorghum in the Luangua valley of the eastern province of Zambia. *Proceedings of the 3rd regular workshop of sorghum and millet for southern Africa. Lusake, Zambia*. SADCC/ICRISAT, Bulawayo, Zimbabwe pp 92–107.

Bridgemohan, P. and Brathwaite, R.A.I. (1989) Weed management strategies for the control of *Rottboellia cochinchinensis* in maize in Trinidad. *Weed Research* 29, 433–40.

Brook, R.M. (1989) Review of literature on *Imperata cylindrica* with particular reference to South East Asia. *Tropical Pest Management* 35 (1), 12–25.

Cavers, P.B. and Bough, M.A. (1985) Proso millet (*Panicum miliaceum*) a crop and a weed. In: White, J. (ed.) *Studies on Plant Demography*. Academic Press, pp143–55.

*Chacon, J.C. and Gliessman, S.R. (1982) Use of 'non-weed' concept in traditional agroecosystems of south-eastern Mexico. *Agroecosystem* 8, 1–11.

*Chabrier, C. (1986) *Contribution à l'étude des herbicides au Nord Cameroun: le cas des cultures vivrières encadrées par la société de développement du coton au Cameroun*. DAT CNEARC-ENSAM, Montpellier, 42pp.

Darmency, H., Jusuf, M., Nguyen Van, E., Poirier-Hamon, S., Barreneche, T. and Pernes, J. (1984) Relations génétiques dans le complexe *Setaria viridis. Comptes rendus du 7éme colloque international sur l'écologie, la biologie et al systématique des mauvaises herbes.* Columa, Paris, pp235–42.

Dawson, J.H. and Burn, V.F. (1975) Longevity of barnyard-grass, green foxtail and yellow foxtail seeds in soil. *Weed Science* 23, 437–41.

Deat, M. (1986) La situation de la malherbologie et de la lutte raisonnée contre les mauvaises herbes en Afrique. *Bulletin Phytosanitaire FAO* 44, 88–99.

*Deuse, J. (1986) Adaptation d'espèces semi-désertiques aux conditions d'irrigation du Sénégal. *Comptes rendus du 5ème colloque international sue l'écologie, la biologie et la systématique des mauvaises herbes.* Columa, Dijon, 359–65.

*Deuse, J. and Guillerm J.L. (1979) *Le désherbage du maïs en Afrique de l'Ouest et les principales adventices*. IRAT-CEPE-AGPM, 17pp.

Deuse, J. and Lavabre, E.M. (1979) Le désherbage des cultures sous les tropiques. *Collection Techniques Agricoles et Productions Tropicales XXVII* Ed. Maisonneuve Larose, Paris 310pp.

Egley, G.H. and Duke, S.O. (1985) Physiology of weed seed dormancy and germination. In: Duke, S.O. (ed.) *Weed Physiology: I. Reproduction and Ecophysiology.* CRC Press, Boca Raton, Florida, USA, pp28–64.

Einhelling, F.A. and Leather, G.R. (1988) Potentials for exploiting allelopathy to enhance crop production. *Journal of Chemical Ecology* 14 (10), 1829–44.

Evans, H.C. (1987) Fungal pathogens of some tropical weeds and the possibilities for biological control. *Biocontrol News and Information* 8 (1), 30pp.

FAO (1982) Weeds in tropical crops: reviews of abstracts *FAO Plant Production and Protection Papers* 32, supp. 1, Rome 63pp.

FAO (1987) *FAO Production Yearbook 40*. Collection FAO statistiques, Rome.

*Fenner, M. (1980) Germination tests on thirty two East African weed species. *Weed Research* 20, 135–8.

Fernandes, F.T. and Schaffert, R.E. (1978) Sorghum in Brazil. In: *Proceedings of the international workshop on sorghum diseases*. ICRISAT, Hyderabad, India, pp15–17.

Fisher, H.H., Lopez, F., Margate, L., Elliot, P. and Burrill, L. (1985) Problems in control of *Rottboellia exaltata* in maize in Bukidnon Province, Mindanao, Philippines. *Weed Research* 25, 93–102.

Fleury, A. (1969) Biologie et écologie de *Cynodon dactylon*. *Revue générale de botanique* 76, 605–30.

*Fontanel, P. (1987a) *Effets sur une végétation adventice du sud Saloum (Sénégal) des fumures minérales, du travail du sol et des précédents culturaux*. IRAT, Montpellier, DSP/87/3, 44pp.

*Fontanel, P. (1987b) *Groupements adventices et facteurs écologiques et culturaux,* *comportements des espèces dans le cycle cultural, efficacité des désherbages paysans et voies d'amélioration au Siné Saloum*. IRAT, Montpellier DSV 7, 60pp.

Forcella, F. and Lindstrom M.J., (1988) Weed seed populations in ridge and conventional tillage. *Weed Science* 36 (4), 500–3.

Frere, M. (1984) Ecological zones and production of sorghum and millet. *Agrometeorology of sorghum and millet in the semi-arid tropics. Proceedings of the international symposium 1982*. ICRISAT, Patancheru, India pp33–9.

Fussell, L.K., Serafini, P.G., Bationo, A. and Klaij, M.C. (1987) Management practices to increase yields and yield stability of pearl millet in Africa. *Proceedings of the international pearl millet workshop*. ICRISAT, Patancheru, India, pp255 – 68.

Garwood, N.C. (1989) Tropical soil seed banks: a review. In: Leck, M.A., Parker, V.T. and Simpson, R.L. (eds) *Ecology of Soil Seed Banks*. Academic Press, pp149–209.

*Gill, S.H. (1986) Rôle du désherbage manuel et mécanique dans la lutte raisonnée contre les mauvaises herbes dans les pays en développement. *Bulletin Phytosanitaire FAO* 44, 20–8.

Grichar, W.J. and Boswell T.E. (1985) Postemergence grass control in peanuts. *Weed Abstracts* 1986, 35 (8), ref. 2721.

*Häfliger, E. and Scholz, H. (1980) *Grass Weeds* 1, 2 Documenta Ciba–Geigy. Basel, Switzerland.

Harinayarana, J. (1987) Pearl millet in Indian agriculture. *Proceedings of the international pearl millet workshop*. ICRISAT, Patancheru, India, pp5–18.

Harlan, J.R. (1982) Relationship between weeds and crop. In: Holzner, W. and Numata, N. (eds) *Biology and Ecology of Weeds*. Dr W. Junk Publishers, The Hague, pp91–6.

*Hashimoto, G. (1982) Brazil. The agrestal weed flora. In: Holzner, W. and Numata, N. (eds) *Biology and Ecology of Weeds* Dr W. Junk Publishers, The Hague, pp333–8.

Hawton, D. and Drennan, D.S. (1980) Studies on the longevity and germination of seeds of *Eleusine indica. Weed Research* 20, 217–23.

Hedge, B.R., Vishwatah, A.D., Reddy, M.N., Satyaranayana, T. and Havanagi, G.V. (1983) Crop weed competition in dryland ragi. *Weed Abstracts* 1986, 35 (7), ref. 3195.

*Holm, L.G., Plucknett, D.L., Pancho, J.V. and Herberger, J.P. (1977) *The World's Worst Weeds. Distribution and Biology*, The University Press of Hawaii, Honolulu, 609pp.

*Ivens, G.W. (1989) *East African Weeds and Their Control* (2nd edn). Oxford University Press, Nairobi, Kenya.

Julien, M.H. (1987) *Biological Control of Weeds; a World Catalogue of Agents and their Target Weeds*. 2nd edn. CAB International, Wallingford, Oxfordshire, UK, 150pp.

*Jurgens, G. and O'Reilly, R. (1975) Control de malezas en mais y sorgo. In: *Curso Basico Sobre Control de Malezas en la Republica Dominicana*. GTZ RFA, pp99–105.

*Kellman, M.C. (1974) The viable seed content of some tropical agricultural soils. *Journal of Applied Ecology* 11, 669–77.

Kigel, J. and Keller, D. (1985) Asexual reproduction of weeds. In: Duke S.O. (ed.) *Weed Physiology: I. Reproduction and Ecology.* CRC Press, Boca Raton, Florida, pp65–100.

Koch, W., Beshir, M.E. and Unterladstatter, R. (1982) Crop losses due to weeds. *Plant Protection Bulletin FAO* 44, 153–65.

*Krantz, J., Schmutterer, H. and Koch, W. (eds) (1977) *Diseases, Pests and Weeds in Tropical Crops*. Verlag Paul Parey. Berlin, 666pp.

*Leger, F. (1990) 'Etude des systèmes sylvopastoraux des montagnes de Colima.' Thèse Univ. Paris VI, 220p.

Lopez, C., Abramovsky, P., Verdier, J.L. and Mamarot, J. (1988) Estimation du stock semencier dans le cadre d'un essai étudiant l'influence de systèmes culturaux sur l'évolution de la flore adventice. *Weed Research* 28, 215–21.

McWhorter, C.G. (1972) Factors affecting Johnsongrass rhizome production and germination. *Weed Science* 20 (1), 41–5.

McWhorter, C.G. and Jordan T.N. (1976) The effect of light and temperature on the growth and development of Johnsongrass. *Weed Science* 24 (1), 88–91.

*Maillet, J. (1991) Place des mauvaises herbes du maïs de riego dans les systèmes agraires à Colima Mexique. In: *Systèmes Agraires de Colima* (in press).

*Marks, M.K. (1983) Periodicity of seedling emergence in six monocotyledonous weeds of south eastern Nigeria. *Oecologia Applicata* 4, 75–85.

*Marks, M.K. and Nwachuku, A.C. (1986) Seed bank characteristics in a group of tropical weeds. *Weed Research* 26, 151–7.

*Marnotte, P. (1984) Influence des facteurs agroécologiques sur le développement des mauvaises herbes en climat tropical humide. *Comptes rendus du 7ème colloque international sur l'écologie, la biologie et la systematique des mauvaises herbes.* Columa, Paris. pp183–9.

Marnotte, P. and Tehia, K.E. (1983) Contribution à la mise au point du désherbage du maïs en Côte d'Ivoire. *Comptes rendus 2ème conférence Société Ouest africaine de Malherbologie*. Abidjan, Ivory Coast, pp226–35.

Marnotte, P and Tehia, K.E. (1989) *Les herbicides dans les cultures vivrières; les produits utilisables*. Note technique IDESSA, Bouakè, Côte d'Ivoire, 16pp.

Matthews, L.J. (1984) Improving weed management at the small-farmer level. *Plant Protection Bulletin FAO* 32 (4), 129–38.

Mattlon, P.J. (1987) Making millet improvement objectives fit client needs: improved genotypes and traditional management systems in Burkina Faso. *Proceedings of the international pearl millet workshop*. ICRISAT, Patancheru, India, pp54–65.

Mbaye Ndoye and Gahukar, R. (1987) Les insectes ravageurs du mil en Afrique de l'Ouest et les moyens de lutte. *Proceedings of the international pearl millet workshop*. ICRISAT, Patancheru, India, pp183–94.

Mercado, B.L. (1978) Biology, problems and control of *Rottboellia exaltata*: a monograph. *Biotrop Bulletin* (14), 38pp.

*Merlier, H. (1972) Etudes phénologiques des espèces de jachère du centre Sénégal. *Agronomie Tropicale* 27 (12), 1229–52.

*Merlier, H. and Montégut, J. (1982) *Adventices Tropicales*. ORSTOM–GERDAT–ENSH Ministère des relations exterieures de la coopération et du développement, Paris, 490pp.

Mersie, W. and Parker, C. (1983) Response of Teff to 2,4D and MCPA at various growth stages. *Weed Research* 23, 53–9.

Michieka, R.W. (1988) Weed control in Kenya. *Comptes rendus du 8ème colloque international sur l'écologie, la biologie et la systématique des mauvaises herbes*. Columa, Paris, pp685–94.

Miller, S.F. (1982) Economics of weed control in the tropics and sub-tropics. In: *Proceedings of the British crop protection conference, Brighton*, pp679–87.

Montillet, J. (1988) *Essais plurilocaux 'désherbage chimique' des cultures vivrières de riz pluvial, maïs, igname et manioc en zone forestière de Côte d'Ivoire*. DAT, CNEARC, Montpellier, 54pp.

Moreira, I. (1978) Influencia da profundidade de enterramento de fragmentos de rhizomas na regeneraçaoces de plantas de *Cynodon dactylon* e de *Panicum repens*. *Revista Ciencias Agrarias* 2, 111–18.

Nieto, J., Brando, M.A. and Gonzalez, J.T. (1968) Critical periods of the crop growth cycle for competition from weeds. *PANS* (C) 14, 159–66.

Noda, K., Chaiwiratnukul, L., Kanjanagirawong, S. and Teerawatsakul, M. (1985a) Some biological characteristics of *Pennisetum* spp. in Thailand. *Proceedings of the 10th conference of the Asian Pacific Weed Science Society.* Thailand, pp75–80.

*Noda, K., Teerawatsakul, M., Prakongvong, C. and Chaiwiratnukul, L. (1985b) *Major Weeds in Thailand.* National Weed Science Research Institute project. Bangkok, 142pp.

Parker, C. and Fryer, J.D. (1975) Lutte contre les mauvaises herbes occasionnant d'importantes réductions des ressources alimentaires mondiales. *Bulletin Phytosanitaire FAO* 23, (3/4), 84–98.

Patterson, D.T. (1980) Shading effects on growth and partitioning of plant biomass in cogongrass (*Imperata cylindrica*) from shaded and exposed habitats. *Weed Science* 28, 735–40.

Patterson, D.T. (1985) Comparative ecophysiology of weeds and crops. In: Duke, S.O. (ed.) *Weed Physiology I. Reproduction and Ecology.* CRS Press, Boca Raton, Florida, USA, pp101–30.

*Popay, A.I. and Ivens, G.W. (1982) East Africa. The agrestal weed flora. In: Holzner, W. and Numata, N. (eds) *Biology and Ecology of Weeds*. Dr. W. Junk Publishers, The Hague. pp345–74.

Purvis, C.E., Jessop, R.S. and Lovett, J.V. (1985) Selective regulation of germination and growth of annual weeds by crop residues. *Weed Research* 25, 415–21.

Putnam, A.R. and Duke, S.O. (1978) Allelopathy in agroecosystems. *Annual Review of Phytopathology* 16, 431–51.

Ramaiah, K.V. and Parker, C. (1982) Striga and other weeds in sorghum. *Sorghum in the Eighties I. Proceedings of the international symposium on sorghum*. ICRISAT, Hyderabad, India, pp291–302.

Rao, M.R., Shetty, S.V.R., Reddy, S.L.N. and Sharma, M.M. (1985) Alternate weed management strategies in improved cropping systems on vertisols in semi-arid India. *Proceedings of the 10th conference of the Asian Pacific Weed Science Society*, Thailand, pp81–9.

Reid, P. and Silawve, M. (1987) Finger millet in northern province of Zambia: some aspects of small farmers production and ideas for development. In: *Proceedings 3rd workshop on sorghum and millets for southern Africa. Lusake, Zambia*. SADCC/ICRISAT Bulawayo, Zimbabwe pp134–47.

Rochecouste, E. (1962) Studies on the biotypes of *Cynodon dactylon.* I. Botanical investigations. *Weed Research* 2, 1–23.

*Saxena, K.G. and Ramakrishnan, P.S. (1984) Herbaceous vegetation development and weed potential in slash and burn agriculture (jhum) in N.E. India. *Weed Research* 24, 135–42.

*Schmid, W., Dossekou, M., Koch, W. and Walter, H. (1983) Aspects de modification et de nuisibilité de la flore adventice dans le système de production agricole du Togo. *Proceedings of the 2nd biannual conference of the West African Weed Science Society*. Abidjan, Ivory Coast, pp19–33.

Schweizer, E.E. and Zimdahl, R.L. (1984) Weed seed decline in irrigated soil after six years of continuous corn and herbicides. *Weed Science* 32, 76–83.

Schweizer, E.E, Zimdahl, R.L. and Mickelson, R.H. (1989) Weed control in corn as affected by till plant systems and herbicides. *Weed Technology* 3, 162–5.

*Schwerzel, P.J. (1967) Seed production of some common Rhodesian weeds. *PANS* (C) 13, 215–17.

*Schwerzel, P.J. (1970) Weed phenology and lifespan observation. *PANS* 16, 511–15.

Seth, A.K. (1986) Les traitements herbicides dans les pays en développement. *Bulletin Phytosanitaire FAO* 44, 68–80.

Shaw, W.C. (1982) Integrated weed management systems technology for pest management. *Weed Science* suppl. 30, 2–12.

Shetty, H.S. (1987) Biology and epidemiology of downy mildew of pearl millet. *Proceedings of the international pearl millet workshop*. ICRISAT, Patancheru, India, pp147–60.

Shetty, R.V.R. (1986) Situation de la lutte contre les adventices dans les cultures associées en zone tropicale. *Bulletin Phytosanitaire FAO* 44, 177–87.

Shetty, R.V.R. and Krantz B.A. (1980) Weed research at ICRISAT. *Weed Science* 28, 451–3.

Sidibe, O. (1978) Principal sorghum diseases in Niger. *Proceedings of the international workshop on sorghum diseases* ICRISAT, Hyderbad, India, pp15–17.

Spencer, D.S.C. and Sivakumar, M.V.K. (1987) Pearl millet in Africa. In: *Proceedings of the international pearl millet workshop*. ICRISAT, Patancheru, India, pp19–32.

*Swamy, P.S. and Ramakrishnan, P.S. (1988) Ecological implications of traditional weedings and other imposed weeding regimes under slash and burn agriculture (jhum) in northeastern India. *Weed Research* 28, 127–36.

Tandon, J.P (1985) Wheat improvement programs for the hottest parts of India. *Proceedings of the international symposium at CIMMYT* pp63–7.

Thomas, P.E.L. (1969) Effects of desiccation and temperature on survival of *Cyperus esculentus* tubers and *Cynodon dactylon* rhizomes. *Weed Research* 9, 1 –8.

Thomas, P.E.L. and Allison, J.C.S. (1975) Competition between maize and *Rottboellia exaltata. Journal of Agricultural Science* Cambridge 84, 305–12.

Thurston, H.D. (1984) *Tropical Plant Diseases.* American Phytopathologist Society. St Paul, Minnesota, 208pp.

*Traore, H. (1987) Influence des techniques culturales sur le développement des mauvaises herbes en station et en milieu naturel rural de Côte d'Ivoire. *DEA Botanique tropicale* USTL, Montpellier, France, 50pp.

*Unamma, R.P.A., Ene, L.S.O., Durukwe, S.O. and Enyinnia, T.E. (1986) Integrated weed management for cassava intercropped with maize. *Weed Research* 26, 9–17.

Walker, R.H. and Buchanan, G.A. (1982) Crop manipulation in integrated weed management systems. *Weed Science* supplement 30, 17–24.

Wall, P.C. (1986) Rôle de l'amélioration des plantes dans les pays en développement. *Bulletin Phytosanitaire FAO* 44, 53–67.

Wellhausen, E.J. (1970) The urgency of accelerating production on small-farms. In: *Strategies for Increasing Agricultural Production on Small Holdings*. Myren, D.T. (ed.) CIMMYT Mexico, pp5–9.

de Wet, J.M.J. (1987) Pearl millet (*Pennisetum glaucum*) in Africa and in India. *Proceedings of the international pearl millet workshop*. ICRISAT, Patancheru, India pp3–4.

*William, R.D. and Chiang, M.Y. (1980) Weed management in Asian vegetable cropping system. *Weed Science* 28 (4), 445–51.

Williams, R.J. and Andrew, D.J. (1983) Breeding for resistance and pest resistance in pearl millet. *Plant Protection Bulletin FAO* 31 (4), 136–58.

*Zimdahl, R.L. (1980) *Weed-crop competitions: a review*. International plant protection center, Oregon State University, Corvallis, Oregon, USA, 196pp.

Zimdahl, R.L. (1984) Improving chemical systems of weed control. *Plant Protection Bulletin FAO* 32 (3), 105–10.

Zimdahl, R.L., Moody, K., Lubigan, R.T. and Castin, E.M. (1988) Patterns of weed emergence in tropical soil. *Weed Science* 36 (5), 603–8.

Chapter 9
Control of Weeds in Tropical Grasslands and Forage Crops

L.'t Mannetje

Department of Field Crops and Grassland Science,
333 Haarweg, 6709 RZ Wageningen, The Netherlands

Introduction

Grasslands and forage crops are ecosystems which produce herbaceous feed (forage) for animals. The vegetation of grassland ecosystems is permanent and may consist of grasses, legumes, other herbs, shrubs and trees. Pure grasslands are treeless; vegetation types with less than 40% tree cover and with a continuous grassy layer are called open woodlands or savannas.

Grasslands may be natural or man-made. Natural grasslands have originated and are maintained by climatic and edaphic forces, natural grazing and fire (climax vegetation). The treeless natural grasslands in tropical and subtropical regions are the velds of Africa, the pampas in Argentina and the downs in Australia. The natural tropical savannas are extensively distributed in northern Australia, sub-Saharan Africa and Latin America. Man-made grasslands are established after cultivation of natural grasslands, tree felling or after a phase of arable cropping. Forage crops can be annually sown or consist of perennial grasses or legumes.

Grassland in tropical and subtropical regions represents a very important form of land use for four main reasons. First it covers 23% of the land surface and most of the feed of domestic herbivores is derived from grasslands. This livestock produces all the beef and sheep meats, milk, wool, some other fibres and manure, which is used to fertilize crops or used as fuel in arid and semi-arid regions. Secondly, land in tropical and subtropical regions is usually only used as grassland if it is not suitable for other purposes, either because the climate is too dry, the soil too shallow or infertile or the topography too steep. Thus, if such land were not used as grassland for livestock, it would not contribute to the production of useful goods. Thirdly, grassland is used in crop rotations to maintain or restore soil fertility and fourthly, grass is an excellent soil cover, preventing erosion.

It is unfortunate, but not really surprising that many useful pasture grasses are among the most troublesome weeds in arable farming, because the grasses belong to the fourth largest and most widely distributed plant family. Examples of such grasses are *Cynodon dactylon, Cenchrus ciliaris, Chloris gayana, Panicum maximum, Paspalum dilatatum* and *Pennisetum clandestinum.* In addition, grasslands have the same grassy weeds as crops: the best known example is *Imperata cylindrica.*

Weeds in grasslands

Effects of weeds

Weeds in grasslands and forage crops are species which do not contribute to or adversely affect the net productivity (yield and feeding value) of the vegetation or the productivity and efficient management of the livestock. The effect of weeds may be direct or indirect. Direct effects are the reduction of growth of desirable species through competition for light, nutrients or water, or, in the case of poisonous plants, endangering the production or life of animals. Indirect effects are that weeds can: (i) make the forage less accessible to livestock; (ii) may harbour livestock pests; or (iii) hinder the handling of the animals.

Weeds in grasslands cause great economic losses, not only from reduced yields and nutritive value, but also directly. For example, in Queensland, Australia, the annual wool clip was estimated to be reduced in value by Aus$4 million as a result of contamination by *Xanthium pungens* and annually about 5,000 tonnes of butter is tainted by the weed *Coronopus didymus* (Bailey, 1988). Animal deaths have been caused by the legume genus *Astralagus* in the United States and by *Acacia georginae* and *Gastrolobium grandiflorum* in Queensland (McEwan, 1964; Everist quoted by Bailey, 1988).

Types of grassland weeds

This chapter reviews the weed problem in grasslands as a whole and therefore, not only grassy weeds, but also dicotyletonous species will be considered. Grassland weeds can be divided into the following types:

1. Very thorny shrubs, trees or cacti (e.g. *Acacia hockii* in the East African highlands, *Mimosa pudica* and *M. invisa* in southeast Asia, *Opuntia*

spp. and *Eriocereus* spp. in northern Australia);
2. Unpalatable grasses and other herbs (e.g. *Imperata cylindrica, Paspalum conjugatum*);
3. Poisonous plants (e.g. legumes belonging to the genera *Astralagus, Acacia, Crotalaria* and *Indigofera*; *Lantana* spp.; species containing oxalates);
4. Plants which harm the animals (e.g. *Eleusine jaegeri*, with highly abrasive leaves that cause soreness of the mouth and wearing away of teeth and *Heteropogon contortus* with sharp awns that penetrate the skin of sheep);
5. Plants with burrs (e.g. *Xanthium pungens*) that spoil the wool of sheep;
6. Thickets of woody plants that foster tsetse flies;
7. Thickets of woody plants that hinder mustering.

A species may be valuable in one forage situation, but a weed in another. An example of this is *Heteropogon contortus*, the dominating grass in unimproved savanna in northern Australia. It is a valued species for cattle, but a noxious one for sheep and a weed in sown grass–legume pastures. Other commonly useful species may at times become toxic to animals, e.g. *Setaria sphacelata* and *Cenchrus ciliaris*, two widely used grasses in improved grasslands, may accumulate 3–5% and 1–2% of soluble oxalates, respectively, which can lead to reduced availability of calcium in ruminants (McKenzie and Schulz, 1983). Species of the genus *Brachiaria* have been very widely used for grassland establishment, particularly in Latin America, but can cause photosensitivity in ruminants, caused by the mycotoxin sporidesmin, which is produced by the fungus *Pithomyces chartarum* living on the grass (CIAT, 1981).

The cause of weeds in grassland

There are four main causes of weediness in grasslands, namely overgrazing, inadequate or excessive soil fertility and the introduction of certain exotic species. Overgrazing, particularly in dry regions, leads first to the ingress of annual grasses but can also lead to the invasion of dicotyledonous herbs, shrubs and trees. Low soil fertility in sown grasslands gives rise to the encroachment of native perennial grasses, whereas excessive soil fertility may lead to increases in annual dicotyledonous weeds.

Weeds are usually native or naturalized species, but can also be introduced accidentally, or for a specific purpose. Examples of this category of weeds are the *Opuntia* spp. which covered much of northern Australia in the 1920s and *Parthenium hysterophorus, Eriocereus martinii* and *Lantana camara* which are troublesome weeds in that region now.

There are three phases in the evolution of man-made grasslands:

1. The establishment phase, which is the period between preparing the land for sowing or planting the grassland or forage crop and the start of full production;
2. The productive phase, which is the time span in which the grassland is in full production;
3. The degradation phase, which is not inevitable, but will occur when the grassland has become less productive as a result of improper management.

The last two phases also occur in natural grasslands.

Weeds can play an important role in the establishment phase, should be unimportant and of minor consequence in the productive phase, but dominate the vegetation in the degradation phase. Weeds in permanent grasslands are usually not the cause, but the visible symptoms and result of degradation.

Weeds in grasslands and forage crops should be avoided, but when they do occur, there are several means of controlling them, which can be divided into mechanical, biological and chemical measures. Mechanical methods consist of cultivation and pulling or crushing of woody weeds, and fire. Biological control is exerted by means of grazing management, smothering by other plants and insects. Chemical control is carried out by means of herbicides.

In order to be able to control weeds it is necessary to know how they originated. Michael (1970) presented an overview of weeds in Australian grasslands and Bailey (1988) dealt with weed control in tropical pastures.

Weeds in the establishment phase

The most critical time with regard to weeds in the establishment phase is the time immediately after germination of the sown species. Weeds originate from remnants of the previous vegetation, from the germination of seeds present in the soil seed reserve or from seeds carried by animals or wind. These may be annuals or perennials. They can be controlled by thorough cultivation combined with chemical treatment before sowing, by pre- and post-emergence herbicides and by management that provides optimal growing conditions for the sown or planted species, such as correct nutrition and early mowing or grazing of the grassland. It may involve a mowing treatment to prevent flowering and seed production of the weeds or removal of the weeds manually or chemically. In the case of low-growing weeds on the one hand, or tall-growing or scrambling grassland plants on the other, it is often possible to smother the weeds by allowing the vegetation to develop undisturbed.

Regrowth and suckering of woody weeds in newly planted grasslands on land recently cleared of forest can be controlled mechanically by the so-called 'ball and chain' method, whereby a heavy chain with a large metal ball in the middle is pulled by two bulldozers, thus pulling out and crushing trees. Another mechanical device is the 'roller-chopper', a heavy cylinder with metal ridges, pulled by a tractor, that crushes woody regrowth. Allowing the land to accumulate enough fuel for a hot fire can also be effective.

Woody weeds and cacti can be controlled by the application of a mixture of picloram and 2,4,5-T to individual plants, although recently other chemicals have been found to be equally as effective, with less danger to the environment than 2,4,5-T (Scanlan, 1984; Scanlan and Fossett, 1984).

Grazing by goats can also contribute to the control of woody species, as these herbivores will often prefer the young leaves and twigs of shrubs and trees to grass (Van Rensburg and Hornby, 1948; Skerman, 1953).

A period of arable cropping, or shallow ploughing after clearing, are good ways to prevent regrowth or suckering problems (Coaldrake, 1967). Effective herbicides during the grassland establishment phase mentioned in the literature are glyphosate, paraquat, oryzalin (for *Leucaena leucocephala* establishment) and trifluralin.

Weed problems are increased in the case of grassland species that are slow growing after germination (e.g. *Leucaena leucocephala, Stylosanthes scabra*). Establishment in rows with inter-row cultivation may be practised in such cases, followed by the sowing of grass when the legumes are well established.

As forage crops are comparable to the establishment phase of permanent grasslands, the weed problem is also comparable.

Weeds in the productive phase

Weeds in the productive phase occur when the sward has weakened to allow intruders to take hold. This may be caused by damage to the sward as a result of poaching, wheel marks or the activities of burrowing animals. Other reasons may be inadequate nutrition of the grassland, excessive fertility, under- and overgrazing, unseasonal rainfall, diseases, pests or drought.

Perennial weeds are usually native or naturalized species which are adapted to the climate and to low soil fertility conditions, whereas sown grassland species are often introduced from elsewhere and require a higher level of nutrition to persist and compete with the weeds. An adequate nitrogen nutrition of introduced grasses often determines their persistence (Mannetje and Shaw, 1972, Henzell *et al.*, 1975). When the nitrogen supply is dependent on legumes, the productivity of the grassland can be jeopardized when the legume has become weak or has disappeared (Mannetje and Jones, 1990). Legume persistence is dependent on the ability of the plants to flower and set seed. The vigour of legume plants is adversely affected by drought, low nutrition (particularly of phosphorous and trace elements such as molybdenum), too frequent cutting, overgrazing and the occurrence of pests and diseases. Once the legume component has begun to weaken, the grazing pressure on the remaining legume biomass is increased and unless the stocking rate is reduced, the legume will come under increasing pressure and can disappear rapidly, resulting in inadequate nitrogen supply to the grass and increasing the chances for weed invasion.

So long as the legume component has not deteriorated too much, it can be restored by judicious grazing management, involving a rest period during the main period of flowering and seed production (Jones, 1988).

It may seem incongruous, but weeds can also become important at high soil fertility levels or when unseasonal rain occurs. When high nitrogen levels have developed as a result of good legume growth, or heavy fertilization, and for one reason or another the desirable grass component, which normally would absorb the available soil nitrogen, has weakened or disappeared, weeds are bound to appear and may dominate the vegetation. Thistles are common weed species in these circumstances. In subtropical regions, wet periods during the usually dry cool season, when the desirable species are dormant, often give rise to a flush of dicotyledonous annual weeds of temperate origin, particularly when soil nitrogen levels are high, such as in legume-based or nitrogen-fertilized grasslands.

Burning is a natural phenomenon as well as a management tool for unimproved grasslands (particularly open woodlands or savannas) to keep woody plants under control. Natural fires occur as a result of lightning strikes and together with man-made fires they are responsible for the maintenance of natural savannas in Africa, South America and northern Australia. The exclusion of fires would lead to an increase in woody species and the reduction of forage production. Burning as a form of management is usually carried out at the beginning of the wet season. It causes the removal of uneaten herbage, unpalatable species and seedlings of woody species, it stimulates new growth from increased tillering as light reaches the base of the grasses and it kills livestock parasites. Pratt and Knight (1971) used fire to reduce the populations of *Tarchonanthus camphoratus* and *Acacia brevispica* in rangeland in Kenya.

Burning can also encourage certain weed species, however, e.g. species whose seeds are activated to germinate by high temperatures (e.g. *Heteropogon contortus*, Shaw, 1955) and shrubs and trees which possess underground buds that are activated after fire and regrow with more stems than they had before. Other species are fire resistant (e.g. *Imperata cylindrica* and *Lantana camara*).

For fire to be effective, there must be sufficient fuel. This will be the case only in years of adequate rainfall during the previous growing season and provided the grassland has not been overgrazed. Overgrazing combined with a sequence of dry years is the main reason for woody weed dominance in grasslands.

The most famous example of biological control of weeds is the success of the moth *Cactoblastis cactorum* in completely controlling about 24 million ha of *Opuntia* spp. in northern Australian grasslands and croplands in the 1920s (Dodd, 1926).

Grassy weeds can in some cases be controlled by legumes. For example, *Imperata cylindrica* is relatively intolerant to shading and a vigorous legume, such as *Centrosema pubescens,* can smother it successfully (Chadhokar, 1977). Adequate fertilization and proper management are then necessary to provide the legume with a competitive advantage.

Weeds in the degradation phase

Grassland in the degradation phase is by definition dominated by weeds. The only way to control these weeds and to return the grassland to its productive phase is by a combination of management techniques aimed at removing the weeds and restoring the desirable species. There are several ways to achieve this: a) complete cultivation and resowing or replanting; b) removing or killing the weeds and oversowing with desired species; c) oversowing with trailing legumes that can smother the weeds; and d) burning the weedy vegetation after accumulation of dry matter to provide fuel for a hot fire, followed by sowing desired species in the ashes.

Conclusion

Since bad management is the main cause of weed invasion and degradation of grassland, it should be obvious that any form of weed control is ineffective unless the management is changed to prevent a recurrence of weed invasion.

Weed control is not practised to a large extent in tropical grasslands (except for woody weeds), for a variety of reasons. First, tropical grasslands are usually unimproved and part of extensive production systems, with often low economic value of the products, or used for subsistence. Secondly, most of the grazing lands in Africa are communally grazed, leading to overgrazing in overpopulated areas and without any specific responsibility for the maintenance of the land. Any weed control that would take place would need to be accompanied by a change in management, which will not happen unless there is a change in land tenure with controlled use. Thirdly, the control of herbaceous weeds by chemicals in improved, legume-based grasslands is difficult because the herbicides not only damage the weeds, but also the desired grasses and legumes. The best prospects are for hand-weeding, spot application of herbicides and woody weed control in grasslands, for which extensive research has been carried out.

The general conclusion is that weeds in grasslands can to a large extent be avoided by good management consisting of appropriate fertilization and use, particularly by avoiding under- and overgrazing. Exceptions are the outbreak of disease, pest damage, floods or drought. It is also worth noting that weeds in forage crops can be treated in the same manner as weeds in other field crops.

References

Bailey, D.R. (1988) Weed control in tropical pastures. In: Skerman, P.J., Cameron, D.G. and Riveros, F. (eds) *Tropical Forage Legumes.* 2nd edn. FAO, Rome, 136–47.

CIAT (1981) International Centre for Tropical Agriculture. Tropical Pastures Program. *Annual Report 1981,* 271–82.

Chadhokar, P.A. (1977) Establishment of stylo (*Stylosanthes guianensis*) in kunai (*Imperata cylindrica*) pastures and its effect on dry matter yield and animal production in the Markham Valley, Papua New Guinea. *Tropical Grasslands* 11, 263–272.

Coaldrake, J.E. (1967) Depth of ploughing in relation to depth of suckering and soil type in the control of root suckers of brigalow (*Acacia harpophylla*). *Australian Journal of Experimental Agriculture and Animal Husbandry* 7, 523–7.

Dodd, A.P. (1926) The campaign against prickly pear in Australia. *Nature* 117, 625–6.

Henzell, E.F., Peake, D.C.I., Mannetje, L.'t and Stirk, G.B. (1975) Nitrogen response of pasture grasses on duplex soils formed from granite in southern Queensland. *Australian Journal of Experimental Agriculture and Animal Husbandry* 15, 498–507.

Jones, R.M. (1988) The effect of stocking rate on the population dynamics of siratro in siratro (*Macroptilium atropurpureum*) – setaria (*Setaria sphacelata*) pastures in south east Queensland. III. Effects of spelling on restoration of siratro in overgrazed pastures. *Tropical Grasslands* 22, 5–11.

McEwan, T. (1964) Isolation and identification of the toxic principle of *Gastrolobium grandiflorum. Queensland Journal of Agricultural Science* 21, 1–13.

McKenzie, R.A. and Schulz, K. (1983) Confirmation of the presence of calcium oxalate crystals in some tropical grasses. *Journal of Agricultural Science* 100, 249–50.

Mannetje, L.'t and Shaw, N.H. (1972) Nitrogen fertilizer responses of a *Heteropogon contortus* and a *Paspalum plicatulum* pasture in relation to rainfall in central coastal Queensland. *Australian Journal of Experimental Agriculture and Animal Husbandry* 12, 28–35.

Mannetje, L.'t and Jones, R.M. (1990) Pasture and animal productivity of buffel grass with siratro, lucerne or nitrogen fertilizer. *Tropical Grasslands* 24, 269–81.

Michael, P.W. (1970) Weeds in grasslands. In: Milton Moore, R. (ed.) *Australian Grasslands.* Australian National University Press, Canberra, Australia.

Pratt, D.J. and Knight, J. (1971) Bush control studies in the drier areas of Kenya. Effect of controlled burning and grazing management on *Tarchonanthus/Acacia* thicket. *Journal of Applied Ecology* 8, 217–37.

Scanlan, J.C. (1984) Herbicidal control of woody weeds in central Queensland. 1. Brigalow (*Acacia harpophylla*). *Tropical Grasslands* 18, 26–32.

Scanlan, J.C. and Fossett, G.W. (1984) Herbicidal control of woody weeds in central Queensland. 2. Dawson gum (*Eucalyptus cambageana*), bitterbark (*Alstonia constricta*) and yellowwood (*Terminalia oblongata*). *Tropical Grasslands* 18, 33–8.

Shaw, N.H. (1955) Bunch spear grass dominance in burnt pastures in south eastern Queensland. *Australian Journal of Agricultural Research* 8, 525–34.

Skerman, P.J. (1953) The brigalow country and its importance to Queensland. *Journal of the Australian Institute of Agricultural Science* 19, 167–76.

Van Rensburg, H.J. and Hornby, H.E. (1948) The place of goats in Tanganyika farming systems *East African Agricultural Journal* 14, 94–8.

Chapter 10
Integrated Control of Grassy Weeds in Mixed Crops

P.J. van Rijn

Royal Tropical Institute, Mauritskade 63, 1092 AD Amsterdam, The Netherlands

Introduction – significance of grassy weeds in agriculture

There are about 250,000 species of flowering plants (angiosperms) in the world, grouped into 450 families. Some 3,000 plants have economic value, and of these about 300 species of plants are used for food or feed. Most of the world's food is produced by some 20 plants species.

Holm *et al.* (1977; 1979) recorded that some 8,000 plants behave as weeds, and that about 250 of these have to be regarded as major weeds. In Table 10.1 the world's worst weeds, registered by the above authors, are listed: 17 of them, subdivided into ten grasses, two sedges and five broad-leaved weeds, are most serious under a wide range of climatic and soil conditions in crops. The most notorious aquatic weed, water hyacinth, belongs to the monocotyledonous family Pontederiaceae.

The title of this book might indicate that the majority of the weeds are grassy types of plants, namely grasses and sedges resembling grasses in their early growth stages. Table 10.1 suggests that this interpretation of the title is correct, and further proof of the validity of this assumption is supplied by the fact that of the terrestrial weeds 23% belong to a single family, namely the grasses, 6% belong to the family of the sedges, 69% are broadleaved weeds divided between more than 20 families, and 2% are ferns and their allies. Of the aquatic weeds, 60% belong to the monocotyledonous families, 30% to broadleaved families, and 10% are ferns (Holm, 1982).

Major problems are experienced with plants closely associated with crops. This group of crop-associated weeds (crop mimicry) includes many grasses such as: wild oats (*Avena fatua* and other *Avena* species) in oats; perennial wild rice (*Oryza longistaminata*) and barnyard grasses (*Echinochloa* spp.) in rice; wild sorghum (*Sorghum halepense*) in

Table 10.1. Main weeds of the world.

Latin name	*Common name*
Cyperus rotundus L.	Nut grass Purple nutsedge
Cynodon dactylon (L.) Pers.	Star grass Bermuda grass Couch grass
Echinochloa crus-galli (L.) Beauv.	Barnyard grass
Echinochloa colonum (L.) Link	Awnless barnyard grass Jungle rice
Eleusine indica (L.) Gaertn.	Goose grass Wild finger millet Crowsfoot grass
Sorghum halepense (L.) Pers.	Johnson grass Aleppo grass
Imperata cylindrica (L.) Raeuschel	Alang-alang Lalang Cogon grass Sword grass
Eichhornia crassipes (Mart.) Solms.	Water hyacinth
Portulaca oleracea L.	Purslane Pigweed
Chenopodium album L.	Goose foot Fat hen
Digitaria sanguinalis (L.) Scop.	Hairy or large crabgrass
Convolvulus arvensis L.	Bindweed Field bindweed
Avena fatua and other members of the wild oat group	Wild oat
Amaranthus hybridus L.	Smooth amaranthus Pigweed
Amaranthus spinosus L.	Spiny pigweed Prickly amaranthus
Cyperus esculentus L.	Yellow nutsedge Water grass Nut grass
Paspalum conjugatum Berg.	Sour paspalum Sour grass

Table 10.1 *Continued*

Latin name	*Common name*
Rottboellia cochinchinensis (Lour) W.D. Clayton	Guinea-fowl grass Itch grass Raoul grass

The above-mentioned weeds occur mainly in the Tropics, with the exception of *Chenopodium album*, *Convolvulus arvensis* and *Avena fatua*.
Source: *The World's Worst Weeds*, by Holm *et al*., 1977.

sorghum; goose grass (*Eleusine indica*) in finger millet (*Eleusine corocana*); and wild cane (*Saccharum spontaneum*) in sugarcane. Many of these plants occur also as weeds in other crops, and some of them are listed in Table 10.1.

Another major category of problem weeds is perennial plants, notably the purple nutsedge (*Cyperus rotundus*) and perennial grasses such as *Cynodon dactylon* (couch or star grass), *Digitaria scalarum* (couch grass), *Imperata cylindrica* (alang-alang), *Paspalum conjugatum* (sour grass), and *Panicum repens* (torpedo grass).

Criteria for selecting mixed cropping systems

A mixture of crops has the ability to suppress weeds better than a monoculture, because the system of intercropping allows only limited space for the growth of weeds, affecting their uptake of water and nutrients, and it provides a more complete canopy preventing germination of light-sensitive weeds such as *Portulaca oleracea*, *Ageratum conyzoides* and *Euphorbia hirta*, and shading out newly germinated and emerged weeds.

Yields of mixed crops are thus increased by weed suppression, but this effect is nullified to a certain extent by the competition between crops. This implies that from an agronomic point of view, intercropping is only considered beneficial if the total production per unit area of the mixture of crops, expressed in yields or in financial returns, is higher than the production of the sole crop. Models for interpreting the mutual interference between plants have been developed by de Wit (1960) and de Wit and van den Bergh (1965). Intercropping is recommended from a soil conservation point of view, because it limits the spatial extent of bare soil, providing cover against the erosive force of water drops, and slowing down surface run off.

Small farmers prefer intercropping to safeguard themselves against crop failure caused by adverse climatic conditions, to reduce the damage caused by pests and diseases, to prevent erosion, and to make more efficient use of labour. On large farms growing annual crops, intercropping is uncommon because of the difficulties experienced with mechanical harvesting.

In many instances suppression of weeds is not considered by the farmers as one of the criteria for selecting mixed cropping. On poor soil farmers often practise wide spacing to enable sufficient uptake of the soil nutrients by the individual plants. However, with wide spacing, weeds are not suppressed and consume large amounts of nutrients and water, resulting in low yields if no intensive weed control is conducted. It is obvious that the farmer has to balance the need for a plant density that reduces the growth of weeds with the fertility status of the soil. In meeting this objective, attention must be paid not only to adjusting plant density, but also to spatial arrangement of crops, and growing of crops with different growth periods and maturity dates. Relay cropping is an efficient means for spreading the demands of the crops.

Although intercropping may contribute markedly to weed control, and efficient uptake of nutrients is assured by selecting proper mixtures of crops, especially by including nitrogen-fixing legumes, it is apparent that additional weed control measures are required, and that fertilization of poor soil is needed to obtain optimum yields.

Major intercropping systems

Crop mixtures vary from one ecological zone to another. In the humid and subhumid tropics major cropping systems are based on rice, maize, cassava, yam, banana/plantain, and newly established plantation crops such as oil palm, coffee, cocoa and rubber. Major cropping systems in the savanna regions are based on stable food crops such as rice, maize, sorghum, pearl millet, finger millet, and on cash crops such as cotton and sugarcane. Some examples of mixed cropping are: maize and beans; maize and cassava and/or yam; relay cropping with quick maturing maize and long-season sorghum; relay cropping of rice and sweet potato. In India where land and water are major limiting factors, intercropping is widely practised in rainfed and irrigated areas, examples being groundnut, mungbean (*Vigna radiata*), urdbean/blackgram (*V. mungo*), or pigeon pea mixed with cotton, and mustard or mungbean as companion crops of sugarcane.

In refining intercropping systems, attention has to be paid to developing early maturing crops that are able to suppress weeds, can withstand drought, and are resistant to some of the major pests and diseases. In India, this type of work receives first priority.

Weed control in mixed cropping

Research on weed control in mixed crops is based primarily on selecting and breeding early maturing, weed competitive crop cultivars. Further research studies of major importance in this field deal with: rotations of a cropping mixture with a sole crop in which mechanical cultivation or weed control by establishing a ground cover are important components; biological control; and chemical control of hard-to-kill weeds, especially perennial grasses.

The practice of intercropping generally prevents the possibility of mechanical weed control during the growing period, because of the limited space available for operating the equipment, but weed control by mechanical means is a most important factor during land preparation.

Weed control in mixtures of annual crops

Intercropping of maize and other cereals with low-growing crops such as cowpea, mungbean and other beans, sweet potato and melon (*Colocynthis citrullus*) is effective in smothering many weeds (Bantilan and Harwood, 1973; Bantilan *et al.*, 1974; Moody, 1980; Akobundu, 1980a; Wahua, 1985). Best results are obtained if the crops, especially the 'smother' crops, are planted immediately after land preparation, so that their canopy can control the first flush of weeds. Shetty and Rao (1981) reported that cowpea or mungbean planted in a sorghum/pigeon pea intercrop was so effective in smothering weeds that it could replace one hand-weeding without affecting the yield of the main intercrops. Akobundu (1987) recorded that sweet potato and melon are good substitutes for repeated hand-weeding in maize/cassava, maize/yam, and maize/yam/cassava intercropping systems. Andrews (1972) obtained good weed control by alternating single rows of late-maturing sorghum with double rows of early-maturing millet followed by cowpea. At IITA in Nigeria good weed control has been obtained by rotating the low-growing crops, melon and sweet potato. The sweet potatoes are planted at the time the melons die off, and in this way a permanent ground cover is maintained by the crops (IITA, 1979).

In addition to selecting low-growing crops, breeding of cultivars that are early-branching or profusely tillering is of major importance from a weed control point of view. Akobundu (1980b) noted that early-branching cultivars of cassava are bred which form a closed canopy in 10–12 weeks, and which thus require less hand-weeding and lower rates of herbicides than the late-branching, local varieties, and the non-branching improved cultivars. Improved dwarf rice cultivars with good tillering ability compete better with weeds than cultivars with moderate tillering (Akobundu and Ahissou, 1984).

Annual rotations and sequential cropping systems (in areas with bimodal rainfall) contribute to maintaining soil fertility (especially if nitrogen-fixing legumes are used), and prevent a build-up of pests, diseases and weeds. Rotations are, among others, needed for control of weeds that are closely associated with certain crops, such as the parasitic weeds, for example, *Striga* spp., which particularly affect cereals. For control of weeds not suppressed by mixed crops, or by sole crops plus ground covers, rotations are recommended with sole crops in which mechanical weed control can easily be practised. Systemic crop rotations are not common in the humid and subhimid tropics, and are also limited in the tropical savanna regions. Like intercropping, crop sequence is determined by the farmer's need for certain species, and by the soil fertility (Akobundu, 1987). The above implies that there is a need for developing specific rotations and crop sequences contributing to the reduction of weeds, pests and diseases. Weeds not controlled by the intercropping and rotation systems are commonly removed by hand. Chemical weed control in mixed crops is practised to some extent. Akobundu (1987) included in his book a list of herbicides reported to be safe in certain cropping systems. Among others, he quoted that: bentazon is safe in maize intercropped with rice, beans, soyabean, or groundnut; pendimethalin in maize/groundnut/beans; metolachlor plus metobromuron in maize intercropped with beans, groundnut, cassava or yam; atrazine plus metolachlor in maize/cassava/yam; and CDEC in maize/vegetables. However, the choice of herbicides for weed control in the various intercropping systems is still limited, and this implies that multi-crop herbicide screening has to be intensified to obtain insights into crop tolerance and herbicide persistence.

Biological weed control, with insects or with microbes such as fungi (mycoherbicides), has not yet been examined for application in mixed cropping systems. Release of insects has been very useful in eliminating the dominance of certain plants in uncultivated fields and waterways, examples being: control of prickly pear (*Opuntia* spp.) by the moth *Cactoblastis cactorum* and the cochineal insect *Dactylopius ceylonicus*; and the control of the aquatic alligator weed (*Alternanthera philoxeroides*) by the flea beetle *Agasicles hygrophila* (Goeden, 1978; Julien, 1982). Insects keep explosions of plant populations in check, and help maintain a certain level of the plants they feed upon. This equilibrium between plant and insect populations cannot be reached in annual and seasonal crops where weeds are destroyed each year during land preparation. Another reason for deciding against release of insects might be that the plant population after reaching equilibrium with the insect population could be too high from a weed control point of view. Naturally, release of insects is not considered at all if their new environment is not free of predators, and if they could become a pest on agricultural crops.

Microbes (fungi and bacteria) are more promising than insects for weed control in annual cropping systems, because of the occurrence of more microbes with host specificity, and the ease of developing/multiplying vast amounts of pathogens/microbes in controlled environments, which are applied to the target as mycoherbicides. In Australia, control of skeleton weed (*Chondrilla juncea*) in wheat was obtained by releasing the rust *Puccinia chondrillina*, which is indigenous in the Mediterranean region (Cullen, 1976). In Arkansas, USA, the endemic/locally occurring fungus *Colletotrichum gloeosporioides* was multiplied *in vitro*, and applied successfully as a mycoherbicide against northern jointvetch (*Aeschynomene virginica*) in rice and soyabean fields (Daniel *et al.*, 1973; Templeton and Smith, 1982). Plants already under stress by climatic conditions or competition are most rapidly killed by biological control agents (Harris, 1981). Research is recommended on development of host-specific mycoherbicides for control of hard-to-kill weeds which can withstand the competitive effects of mixed crops and ground covers. Mycoherbicides could replace or supplement herbicides for control of nutsedge and dominant perennial grasses, especially alang-alang, *Paspalum conjugatum* (sour grass), *Cynodon dactylon* (couch grass) and *Panicum repens* (torpedo grass).

To obtain satisfactory weed control in a mixture of annual crops, some or all of the methods described above have to be applied. The age-old farmer's practice of mixed cropping plus hand-weeding is the basis of integrated weed management. The value of this integrated system can be improved by refining intercropping and rotation systems, and by partly replacing hand-weeding by chemical or biological applications.

Weed control in young plantations interplanted with annual crops

Intercropping is conducted during the early lifetime of the tree crop until it comes into bearing (rubber, oil palm), or before it has formed a closed canopy (coffee, tea). In coconut, intercropping can be carried out during the whole lifetime of the crop, but food crops such as maize, beans and bananas are particularly grown during the period when the coconut is not bringing in any financial return.

The use of ground covers is a well-known, established practice in tree crops in many tropical countries for preventing soil erosion, preserving soil structure, regulating soil fertility, and preventing, suppressing or controlling weeds through competition for light, water and nutrients. Intercropping controls weeds to a certain extent, and the tree crops receive maximum benefit from this practice if additional weed control measures are conducted in the secondary crops. Ground covers can be used for control of weeds both in perennial and annual crops. Akobundu (1980c) found that maize planted in fields with the legumes centro (*Centrosema pubescens*) and wild winged bean (*Psophocarpus palustris*) gave very good weed control without affecting the yield of the crop. Main ground covers include the creeping legumes *Centrosema pubescens*, *Indigofera spicata*, *Pueraria phaseoloides* and *Calopogonium mucunoides*, and the leguminous shrubs *Crotalaria* spp., *Tephrosia* spp., and *Moghania macrophylla* (*Flemingia congesta*).

The weeds in cocoa, coffee and tea can also be ecologically controlled by growing shade trees, notably *Albizia* spp., *Gliricidia* spp., *Erythrina* spp. and *Leucaena* spp. Cocoa should be given adequate shade until it becomes self-shading or permanent overhead shade has become established. Bananas or plantains planted before cocoa are most frequently used as temporary shade. Other plants used for this purpose include *Crotalaria* spp., *Tephrosia* spp., papaya, *Manihot glaziovii* (tree cassava), *Manihot esculenta* (cassava) and *Cajanus cajan* (pigeon pea). Young cocoa trees are very susceptible to weed competition, and this implies that: ringweeding by hand or with chemicals has to be conducted; weeds/ground covers in the remaining part of the cocoa plantation have to be removed in the third year; and temporary overhead shade such as from banana has to be gradually removed in the fourth year (Purseglove, 1968). Coffee is by nature suited for growing without shade, but despite this it is grown under shade in most areas, because: i) shade may protect coffee bushes from excessive insolation; ii) shade trees produce large amounts of organic matter that contribute to a good physical condition of the soil and render nutrients available from the subsoil in the form of leaf-fall; and iii) shade and mulch from shade trees suppress weeds. In Brazil and Kenya coffee is grown without shade, because rainfall is too low to provide sufficient moisture for both coffee and shade trees. In the case of tea, shade has proved beneficial for light-leaved types with large, horizontal leaves in low-altitude/high temperature areas in the tropics.

Ringweeding and weeding of the tree crop rows are usually done by hand-pulling, hoeing and slashing, but chemical weed control is not uncommon. Notorious weeds surviving ground covers include: the broad-leaved plants *Chromolaena odorata* (*Eupatorium odoratum*), *Melastoma malabathricum*, *Melampodium* spp., *Galinsoga parviflora*, *Bidens pilosa* and the 'escaped' ground covers *Mikania* spp., and *Mimosa* spp.; the sedge *Cyperus rotundus*; and the grasses *Imperata cylindrica*, *Cynodon dactylon*, *Paspalum conjugatum*, *Panicum repens*, *Digitaria scalarum* and *Axonopus compressus*. Herbicides such a metolachlor and oxyfluorfen control a range of annual weeds during legume cover establishment. Non-selective herbicides such as glyphosate and paraquat are used as directed sprays for control of hard-to-kill, mainly perennial weeds (spot treatments). Development of mycoherbicides for control of specific weeds in ground covers is recommended.

Dead mulches such as leaves shed by shade trees, crop residues, and grasses procured from pastures (e.g. *Pennisetum* or *Panicum* spp. grown and harvested for this purpose) are used to prevent germination of light-sensitive weeds, and to suppress weeds by light interception and by allelopathy, that is the release of toxic materials. Toxins can also be exuded by roots and other parts of the living plant, some examples being: the deleterious effect of the ground cover *Salvia occidentalis* on young coffee trees (van der Veen, 1935), and the inhibitory effect of alang-alang on growth of maize (Eussen *et al.*, 1976). Despite the examples cited above, it must be mentioned that the damage caused by allelopathy from living plants is insignificant compared to the suppression of plants due to competition. However, the role of allelopathy is clearly apparent when dead mulches are used for weed control.

Most efficient weed control can be obtained by combining and refining some or all of the above-described methods. In future, special attention has to be paid to the hard-to-kill weeds, many of which are perennial grasses. Furthermore, emphasis is needed on the development of mycoherbicides.

Concluding remarks

No single weed control method will control sufficiently all weeds in a crop or a mixed of crops. Weeds, especially perennial grasses, form a major problem in mixed cropping systems, which can only be satisfactorily controlled by an integrated system of control methods, consisting of a combination of some or all of the following measures:

- cultural treatments (density and time of planting, use of dead mulches);
- ecological treatments (systems of intercropping and rotations, and use of ground covers/living mulches);
- hand-weeding (pulling, hoeing and slashing);
- chemical treatments;
- biological treatments, especially the use of microbiological weed control agents/mycoherbicides.

References

Akobundu, I.O. (1980a) Weed control strategies for multiple cropping systems of the humid and subhumid tropics. In: Akobundu, I.O. (ed) *Weeds and their Control in the Humid and Subhumid Tropics*, IITA Proceedings Series No. 3, IITA, Ibadan, Nigeria, pp80–107.

Akobundu, I.O. (1980b) Weed control in cassava cultivation in the subhumid tropics. *Tropical Pest Management* 26, 420–6.

Akobundu, I.O. (1980c) Live mulch: a new approach to weed control and crop production in the tropics. Proceedings of the 1980 British Crop Protection conference. *Weeds* 2, 377–82.

Akobundu, I.O. (1987) *Weed Science in the Tropics – Principles and Practices.* John Wiley & Sons, Chichester/New York, XVI, 522pp.

Akobundu, I.O., and Ahissou, A. (1984) Effect of interrow spacing and weeding frequency on the performance of selected rice cultivars on hydromorphic soils of West Africa. *Crop Protection* 4, 71–6.

Andrews, D.J. (1972) Intercropping with sorghum in Nigeria. *Experimental Agriculture* 8, 139–50.

Bantilan, R.T. and Harwood, R.R. (1973) Weed management in intensive cropping systems. Saturday seminar paper, 28 July, 1973, International Rice Research Institute, Los Banos, Laguna, Philippines, 7pp.

Bantilan, R.T., Palada, M.C. and Harwood, R.R. (1974) Integrated weed management. 1. Key factors affecting crop–weed balance. *Philippine Weed Science Bulletin* 1 (2), 14–36.

Cullen, J.M. (1976) Evaluating the success of the programme for the biological control of *Chondrilla juncea.* In: Freeman, T.E. (ed.) *Biological Control of Weeds.* University of Florida Press, Gainesville, Florida, pp117–21.

Daniel, J.T., Templeton, R.J., Smith, R.J. Jr. and Fox, W.T. (1973) Biological control of northern jointvetch in rice with an endemic fungal disease. *Weed Science* 21, 303–7.

Eussen, J.J.H., Slamet, S. and Soeroto, D. (1976) Competition between alang-alang (*Imperata cylindrica* (L.) Beauv.) and some crop plants. *BIOTROP Bulletin* 10, 3–24.

Goeden, R.D. (1978) Biological control of weeds. In: Clausen, C.P. (ed.) *Introduced Parasites and Predators of Arthropod Pests and Weeds: a World Review.* US Department of Agriculture, Agricultural Handbook No. 480, pp357–506.

Harris, P. (1981) Stress as a strategy in the biological control of weeds. In: Papavizas, G.C. (ed.) *Biological Control in Crop Production.* BARC symposium No. 5., Allanheld Osmum, Totowa, New Jersey, pp333–49.

Holm, L.G., Plucknett, D.L., Pancho, J.V. and Herberger, J.P. (1977) *The World's Worst Weeds – Distribution and Biology.* The University Press of Hawaii, Honolulu, 609pp.

Holm, L.G., Pancho J.V., Herberger, J.P. and Plucknett, D.L. (1979) *A Geographical Atlas of World Weeds.* John Wiley & Sons, New York, XLIX, 391pp.

Holm, L.G. (1982) The biology and distribution of some weeds important to the tropics. *Proceedings of the International Conference on Plant Protection in the Tropics,* Malaysian Plant Protection Society, pp85–97.

IITA (1979) *Research and training activities at IITA.* International Institute of Tropical Agriculture, Ibadan, Nigeria, 11, 151pp.

Julien, M.H. (1982) *Biological Control of Weeds: a World Catalogue of Agents and Their Target Weeds.* 1st edn. CAB International, Wallingford, Oxfordshire. 108pp.

Moody, K. (1980) Weed control in intercropping in tropical Asia. In: Akobundu, I.O. (ed.) *Weeds and their Control in the Humid and Subhumid Tropics.* IITA Proceeding Series No. 3, IITA, Ibadan, Nigeria, pp101–8.

Purseglove, J.W. (1968) *Tropical Crops – Dicotyledons.* Longman Group Limited, London, 719pp.

Shetty, S.V.R. and Rao, M.R. (1981) Weed-management studies in sorghum/pigeon pea and pearl millet/groundnut intercrop systems – some observations. In: Willey, R.W. (ed) *Proceedings of the International Workshop on Intercropping*, 10–13 January, 1979, ICRISAT (International Crops Research Institute for the Semi Arid Tropics), Hyderabad, India.

Templeton, G.E. and Smith, R.J. Jr. (1982) The role of plant pathogens in weed management in advancing countries. *FAO Plant Production and Protection Paper* 44, pp85–90.

van der Veen, R. (1935) Wortelconcurrentie in de koffie en rubbertuinen (Root competition in coffee and rubber plantations). *Archief van de Koffiecultures* 9, 65–104.

Wahua, T.A.T. (1985) Effect of melon (*Colocynthis vulgaris*) population density on intercropped maize (*Zea mays*) and melon. *Experimental Agriculture* 21, 281–9.

de Wit, C.T. (1960) On competition. *Verslagen van landbouwkundige onderzoekingen* No. 66, 8, 82pp.

de Wit, C.T. and van den Bergh, J.P. (1965) Competition between herbage plants. *Netherlands Journal of Agricultural Science* 13, 212–21.

Chapter 11
Weed Control in Upland Rice with Emphasis on Grassy Weeds

K. Moody

Agronomist, The International Rice Research Institute, Los Baños, Laguna, Philippines

Summary

All plants are useful and beneficial in some way but some plants are called weeds because they are growing in situations that do not serve man's needs. They can cause losses to farmers by increasing the cost of growing crops or harvesting and they reduce the yield and quality of cultivated crops. Weeds are more of a problem in upland rice (*Oryza sativa* L.) than in lowland rice because in the latter weeds can be controlled to some extent by submergence, provided that there is good control of irrigation water. The most significant management practice which could have the greatest impact on upland rice yields is improved weed control. Grasses are predominant and still remain a serious problem in upland rice culture. Their effects on upland rice and possible control measures are discussed in this chapter.

Introduction

When shifting cultivation is practised, the end of the fallow occurs with land clearing. Clearing is frequently, but not always, followed by burning (Moody, 1982). For example, in the Pespire region in southern Honduras, generally in the first year, bush and small trees are cut down during the middle of the rainy season and are allowed to remain in the field to act as a mulch. The mulch acts as a ground cover to conceal the seeds from potential pests, retains soil moisture to help the seeds germinate and helps to impede weed growth. Burning is implemented at the start of the second year of cultivation (de Walt *et al.*, 1985).

Velasco *et al.* (1961) reported that most of the weeds found in upland rice in the first crop following clearing of the bush fallow were broadleaved weeds but during the second cropping season, the composition of the weed flora shifted mainly to grasses. In Peru the weed population in some upland rice areas has changed from one dominated by broadleaved weeds to one dominated by grasses after two crops (Sanchez and Nurena, 1970).

Moody (1982) reported that during the first crop following forest clearing, broadleaved weeds accounted for 85% of the weed population in the unweeded plots. A dramatic change in the weed flora occurred in the second crop with grasses accounting for 77% of the weed flora in the unweeded plots. The shift towards grasses continued in the third crop.

Based on a survey conducted in 36 countries representing approximately 90% of the upland rice area in the world, weeds are the major biological constraints to rice production in 25 of those countries (10,416,800 ha or 58% of the area) (Arraudeau, 1986). Yield losses can be almost total if there has been no weed control.

Weeding has long been recognized as one of the important determinants of agricultural yield in the tropics. Weed growth is so prolific and so difficult to control that it is the major yield-depressing factor (Ruthenberg, 1980). For most farmers, weeding is the biggest job in the agricultural cycle and it is the step that limits the amount that can be seeded since the first weeding must be completed early in the season before weed competition hurts the crop (Cancian, 1972). Farmers in Bangladesh are limited in the area of land that can be cropped due to the heavy labour demand needed for weeding (Elahi *et al.*, 1985). Often labour is not available for weeding when weeds are most damaging to the crop.

In upland rice, weeds grow faster than in lowland fields and are so hardy that they are difficult to control (Camus, 1921). Weed control is one of the most important problems of upland rice. Weeds increase the cost of production and lower the amount and grade of the rice produced. Their control is difficult when they grow at the same rate as the crop; sometimes when the land is too weedy it is impossible to save the crop.

Weed competition in upland rice reduces plant height, number of shoots/m^2, dry matter production, leaf area index, grain yield, panicle number, panicle weight and panicle length (Gunesena, 1974).

The upland rice production system is highly dynamic with respect to many factors which may significantly alter the environment from year to year. In West Africa, although upland rice is raised in ecological zones from forest through savanna to mid-altitudes, the weed flora in these zones does not appear to vary greatly. What does vary is the intensity of the infestation and the dominant weed species in a particular habitat. *Rottboellia cochinchinensis* (Lour.) W.D. Clayton may be the major weed limiting rice production in one area while *Echinochloa colona* (L.) Link predominates in another (Akobundu and Fagade, 1978).

Grassy weeds predominate in upland rice. There are numerous grassy weed species reported to occur in upland rice throughout the world. For example, Moody (1989) lists about 140 for selected countries in south and southeast Asia. Of these, *Dactyloctenium aegyptium* (L.) Willd., *Digitaria* spp., *Echinochloa colona, Eleusine indica* (L.) Gaertn., *Imperata cylindrica* (L.) Raeuschel, and *R. cochinchinensis* appear to be the most important on a worldwide basis.

Critical period of competition

During the cropping period, there is a particular length of time, the critical period of competition, during which the crop is very sensitive to competition and the presence of weeds above a certain density will cause a significant yield reduction. Weeds should be controlled within this period. The precise time and duration of the critical period depend on many factors such as the weed flora, growth characteristics of the rice and weeds, cultural practices and environmental factors (Moody, 1977).

It is, therefore, difficult to state how long after seeding a crop should be weeded and how many times it should be weeded during the period of greatest competition. However, it is often during the period from two to six weeks after emergence (cf. Anon., 1985; Ghosh *et al.*, 1975; Sahai *et al.*, 1983; Singh *et al.*, 1985) that upland rice yield is lost due to competition from weeds.

Noguchi (1989) reported that the growth of *Digitaria ciliaris* (Retz.) Koel. was greatly reduced by 80–90% shading. The relative light intensity under the canopy of upland rice was reduced to below 10% at about 89 days after seeding (DAS) and at that time the inter-row space up to a height of 30 cm was under the same light intensity. Thirty days were required for *D. ciliaris* to grow to about 30 cm in height. Therefore, a 59 day weed-free period is theoretically required after upland rice is sown in a rice–*D. ciliaris* community.

Weed control

Most cultural practices may be regarded as means of weed suppression and an increase in their efficiency would contribute to better weed control (Anon., 1952).

Burning

The effect of burning on weed seed germination and growth has not been systematically investigated. Akobundu (1983) stated that burning was a low-cost method for eliminating excess vegetation and also destroying weed seeds. However, Crafts and Robbins (1962) noted that, in general, burning removes debris rather than controls weeds.

Fire seldom reduces the viability of weed seeds that have fallen to the soil surface (Crafts and Robbins, 1962) although there are reports that it will destroy all (Clark and Haswell, 1970) or some seeds (Klingman, 1961; Webster and Wilson, 1966; Oyenuga, 1967). Fire may promote weed seed germination by breaking seed dormancy (Went *et al.*, 1952; Klingman, 1961; Crafts and Robbins, 1962) although it has also been reported that burning seemed to delay (Brown, 1969; Ahn, 1970; Hill, 1982) or had no effect (Went *et al.*, 1952) on germination.

Burning, with clearing, will also promote weed seed germination and growth by removing the competition from other plants and the shading effect of litter, allowing more light to reach the soil surface and causing greater temperature alternations in the upper soil layers between day and night (Went *et al.*, 1952; Watters, 1958; Klingman, 1961). It has been observed that burning promoted the germination of *Mucuna pruriens* (L.) DC. in Nigeria and *Mimosa invisa* Mart. in the Philippines.

Land burned too often frequently becomes overgrown with perennial grasses which may render it useless for agricultural production when simple tools are used (Bartlett, 1956). Burning also stimulates the growth of some perennial weeds such as *Imperata cylindrica.* In southeast Asia, *I. cylindrica* covers much of the land that has been lost to the system (Barker and Herdt, 1985).

Land preparation

The success of upland rice depends on the thoroughness of land preparation (Camus, 1921). Poor land preparation is a major factor affecting the severity of weed infestation (Allnut, 1942; Merlier, 1973; Parry, 1974). In India, Biswas and Khan Bassi (1982) reported that weed density and weed weight were greater in a poorly prepared seedbed while in Sri Lanka, Abeyratne (1956) noted that grasses completely swamp the rice crop in fields where tillage operations have not been effective.

Biswas and Khan Bassi (1982) reported that the performance of butachlor [N-(buthoxymethyl)-2-chloro-N-(2,6-diethylphenyl)acetamide] was better in a fine seedbed. Herbicide toxicity was greater in a poorly prepared (cloddy) seedbed because of uneven distribution of the herbicide.

Ploughing depth

Increased tillering, plant height and yield, and better weed control were obtained with deep ploughing of 20–25 cm (Uichanco, 1959). Other authors (Pande and Bhan, 1966; Curfs, 1975; Mesquita, 1988; Singh *et al.*, 1988) have also shown the effectiveness of deep ploughing in reducing weed biomass.

Seeds of *R. cochinchinensis* and *Eleusine indica* may lose their viability if buried deep in the soil (Schwerzel, 1976); *Dactyloctenium aegyptium* can remain viable for at least 6.5 years if buried to a depth of 15 cm (Juliano, 1940).

Time of land preparation

Land preparation at different times of the year influences weed emergence, degree of infestation, and the dominance of a weed species in a community. Janiya and Moody (1979) reported that there was a significant reduction in weed weight at 45 days after emergence (DAE) but not at harvest as a result of land preparation during the dry season compared to land preparation at the start of the rainy season. *Celosia argentea* L. was the dominant weed when land was prepared during the dry season while *R. cochinchinensis* and a *Digitaria* sp. dominated when land preparation was carried out at the start of the rains.

Zero tillage

Upland rice is notoriously high in labour inputs when farmed traditionally (930 h/ha from field preparation to crop maturity). Wijewardene (1980) reduced this to 44 h/ha for a zero tillage system. The time required for weeding was reduced from 500 h/ha for two manual weedings using a hand hoe to 19 h/ha for two herbicide applications. The highest yield was obtained in the zero tillage field.

On a hydromorphic soil in the Guinea savanna, a low rainfall region of Nigeria, Olofintoye (1989) reported that weed biomass was significantly higher in the no-till than in the conventional and minimum tillage treatments. However, herbicides effectively controlled weeds under all tillage practices and were comparable with two hand weedings. Olofintoye (1989) noted that plant height and tillering at early growth stages of the crop were lower in the no-till plots. He thought that some level of soil manipulation was necessary for satisfactory tillering in rice even though the three tillage practices produced similar yield.

Castin *et al.* (1983) concluded that zero tillage cannot be practiced in upland rice until a suitable herbicide technology involving both pre- and post-plant treatments is developed. Pande and Bhan (1964) considered one ploughing with a moldboard plough and one harrowing with a disc harrow to be the minimum requirement for soil manipulation in upland rice.

Cultivar

The widespread replacement of tall, traditional cultivars with semi-dwarf cultivars has increased weed problems in rice. The dwarf cultivars favour more weeds by virtue of having erect leaves that permit light to penetrate deep into the canopy and that also facilitate more air mixing. Furthermore, the high fertilizer rates that are applied to the high-yielding cultivars aggravate weed problems (Anon., 1985).

The taller plant types are more competitive with weeds than the dwarf plant types and are more appropriate for upland conditions (ARS, 1982). BG 62–355, a tall cultivar, was more competitive than BG 34–8, a short cultivar (ARS, 1981). Liboon *et al.* (1978) reported that shorter plant types often intensified weed problems later in the growing season.

When chemical weed control (propanil [*N*-(3,4-dichlorophenyl) propanamide] + MCPA [(4-chloro-2-methylphenoxy)acetic acid]) was used, the yield loss due to weeds with the taller cultivars was 0–3% whereas with the shorter cultivars, it varied from 5–31% (ARS, 1981).

Nitrogen

Sahai and Bhan (1982) reported that there was a significant negative linear relationship between nitrogen uptake by drill-seeded rice and weeds. On the basis of nitrogen uptake, *E. colona* was the most competitive weed species.

Fertilizer application should be timed to prevent weed proliferation and yet obtain maximum benefit for the crop. Fertilizers should be applied immediately after weeding (Dobby, 1955; Selvadurai and Arope, 1969; John and Kalode, 1981). If the weeds are not controlled before the fertilizer is applied then competition increases. If weeds are not adequately controlled, then fertilizer should not be applied (Sankarama, 1986).

The earlier the canopy closes, the lower the loss of nutrients and moisture due to weed competition. A high plant density can suppress weed competition in upland rice (Kawano *et al.*, 1972; Suprapto, 1980; ARS, 1981; Tosh *et al.*, 1981; Kehinde, 1985; Gassaver and Fion, 1988). Simbolon (1985) reported that there was a negative correlation between the species number and soil coverage of weeds with the soil coverage by rice.

Manual weeding

In most of the rice fields of the world, manual weeding is the general practice. One weeding, no matter how well timed, will not provide season-long weed control in upland rice. When upland rice fields are moderately infested with weeds, two weedings are generally sufficient. However, when weed growth is severe, three weedings during the first eight weeks of crop growth may be required to give adequate weed control.

Manual weeding methods can be very effective in controlling weeds but they tend to be extremely time consuming and laborious. Mangunsoekardjo and Kadnan (1971) reported that weeding by hoe in upland rice took 318 h/ha, thus extensive rice cultivation in scarcely populated areas cannot be weeded with a hoe.

Inter-row cultivation

ARS (1981) reported that different inter-row cultivation implements reduced weed weight by only 30–50% because they failed to control weeds within the row. Crop losses of 30–50% occurred in these plots compared to a 10% yield loss when herbicides (propanil plus MCPA) were used for weed control. Singh *et al.* (1988) reported that weed weights were lowest and grain yields were highest when a pre-emergence application of butachlor was followed by inter-row cultivation at 30 DAS.

Herbicides

At present, little or no chemical weed control is used in upland rice. The compounds available are either expensive or unsuitable due to lack of persistence or failure to control a broad spectrum of weeds. Pre-emergence herbicides that have been reported to control weeds effectively in upland rice include butachlor (Kennard, 1973; Ellis and Haws, 1976; De Datta, 1977; Indhapun *et al.*, 1979; Borgohain and Upadhaya, 1980; Sabio *et al.*, 1981), thiobencarb (*S*-[(4-chlorophenylmethyl]diethylcarbamothioate) (Mallick *et al.*, 1974; Bhagat *et al.*, 1977; De Datta, 1977), oxadiazon (3–[2,4–dichloro–5–(1–methylethoxy)phenyl]–5–(1,1–dimethylethyl)–1,3,4-–oxadiazol–2–(3*H*)-one) (Bhagat *et al.*, 1977; De Datta, 1977; Mukhopadhyay *et al.*, 1989), pendimethalin [*N*–(1–ethylpropyl)–3,4–dimethyl–2, 6–dinitrobenzenamine] (De Datta, 1977; Singlachar and Chandrashekar, 1977; Shelke *et al.*, 1986), and a proprietary mixture of piperophos (S–2–methyl–piperidinocarbonyl methyl *0,0*–dipropyl phosphoro–dithioate) and dimethametryn [2-(1,2-dimethylpropylamino)-4-ethylamino-6-methylthio-1,3,5-triazine] (Gopalakrishna Pillai and Rao, 1974; Gunasena, 1974; Modgal and Singh, 1980).

While undoubtedly efficient, herbicides are not miracle solutions. Better results are obtained if pre-emergence herbicide treatments are followed by or combined with a post-emergence application of propanil (Domingo and Palis, 1966; Mukhopadhyay *et al.*, 1971; Renaut, 1972; Aryeetey, 1973; Borgohain and Upadhaya, 1980), manual (Fagade, 1976; Indhapun *et al.*, 1979; Pande *et al.*, 1983; Singh *et al.*, 1983; Dash and Singh, 1984; Ramamoorthy and Ali, 1985; Deshmukh *et al.*, 1987; Gogoi, 1989; Singh and Shrivastava, 1989; Singh and Singh, 1989) or mechanical (Abeyratne, 1956) weeding.

Post-emergence herbicides that have been reported to control grass weeds effectively in upland rice include propanil (Cates, 1967; Mangunsoekardjo and Kadnan, 1971; Singh *et al.*, 1982; Singh and Singh, 1983) and fenoxaprop [(±)-2-[4-[(6-chloro-2-benzoxazolyl)oxy]phenoxy]propanoic acid] (Chavez, 1989). Propanil is used primarily for grass control although it will control some broadleaved weeds and sedges if they are in the seedling stage at the time of treatment. Fenoxaprop only controls grasses.

Better control of predominantly grass weed populations has been achieved with sequential applications of propanil (Mukhopadhyay, 1967; Mayer and Chabrolin, 1972; Tosh *et al.*, 1981) or post-emergence herbicides followed by manual weeding (Mukhopadhyay, 1967; Kolhe and Mittra, 1981; Singh and Dash, 1984; Chavez, 1989).

Intercropping

Sengupta *et al.* (1985) found that blackgram (*Vigna mungo* (L.) Hepper) adversely competed with rice when sown at the same time. When, however, blackgram was sown 21 days after the rice, the competition decreased; rice yield was reduced by only 0.24 t/ha compared with sole cropping of rice. The blackgram reduced the weed population appreciably and eliminated the second weeding required in sole cropped rice. This system provided a monetary advantage of US$116/ha over sole cropping of rice.

In an experiment conducted in Claveria, Misamis Oriental, Philippines, when two upland rice rows spaced 30 cm apart were alternated with one cowpea (*Vigna unguiculata* (L.) Walp.) row (2:1 ratio), cowpea canopy closure after the first weeding operation helped suppress weed growth and reduced the cost of weed control. However, in a similar experiment conducted at IRRI, Los Baños, Laguna, weed biomass at 45 DAE was not suppressed when cowpea and mungbean (*Vigna radiata* (L.) Wilczek) were intercropped with upland rice (Elliot and Moody, unpublished).

Alley cropping

Alley cropping is an improved bush fallow system in which crops, especially food crops, are grown in alleys formed by selected shrub or tree species, usually legumes, established to provide organic residue and recycle nutrients. Depending on the state of the crop of the soil-restoring species, pruning or other methods of suppression may be used to prevent the trees or shrubs from competing with the crop. Weed and erosion control, as well as provision of fuelwood and livestock feed, are also important attributes of the system (Wilson and Kang, 1981).

IITA (1983) reported that the organic residue from *Acioa barteri* Engl. decomposed very slowly and the persistent mulch that it formed prevented weed growth and eliminated the need for weeding cowpea.

Yamoah *et al.* (1986) reported that, during the fallow period, weed control was most effective under *Cassia siamea* Lam. followed by *Gliricidia sepium* (Jacq.) Steud. and *Flemingia congesta* Roxb. Weed biomass during maize (*Zea mays* L.) crop growth was negatively related to the percentage of prunings remaining on the soil surface. *C. siamea*, with the slowest rate of decomposition, had the lowest weed invasion.

Cover crops

In Guatemala, when weeds become especially burdensome, most shifting cultivators, rather than spending days clearing them away prefer to seed velvetbean (*Mucuna deeringiana* (Bort) Merr.) and let it spread over the ground. As it does so, it shades the grasses until they die. Farmers in southeastern Mexico have reduced fallow periods from five years to one growing season by growing a crop of velvetbean and swordbean (*Canavalia gladiata* (Jacq.) DC.) to restore productivity to their fields (NAS, 1979).

In West Africa, *I. cylindrica* is traditionally controlled by hoeing or tillage but these methods do not provide season-long control. In an experiment at IITA, Nigeria, planting of *M. pruriens* var. *utilis* (Wall) Burck. provided excellent cover within 12 weeks and no *I. cylindrica* could be seen above the *M. utilis* cover at 19 weeks. However, the regrowth capability of rhizomes collected from these plots was 50% that of rhizomes from the slashed control plots (IITA, 1984).

References

Abeyratne, E.L.F. (1956) Dry land farming in Ceylon. *Tropical Agriculturist* 112, 191–229.

Ahn, P.M. (1970) *West African Agriculture.* 3rd edn, Vol. 1. *West African Soils.* Oxford University Press.

Akobundu, I.O. (1983) No-tillage weed control in the tropics. In: Akobundu, I.O. and Deutsch, A.E. (eds) *No-tillage Crop Production in the Tropics.* International Plant Protection Center, Oregon State University, Corvallis, Oregon, USA, pp32–44.

Akobundu, I.O. and Fagade, S.O. (1978) Weed problems of Africa rice lands. In: Buddenhagen, I.W. and Persley, G.T. (eds) *Rice in Africa.* Academic Press, London, pp181–92.

Allnut, R.B. (1942) Rice growing in dry areas. *East African Agricultural and Forestry Journal* 8, 103–8.

Anon. (1952) Weed control in paddy. *Tropical Agriculturist* 108, 196–9.

Anon. (1985) Rice research at Pantnagar. *Experimental Station Technical Bulletin* 112, G.B. Pant University Agriculture and Technology, Pantnagar, India, 94pp.

Arraudeau, M. (1986) *Final report on international upland rice genetic survey (IURGS)*. IRRI, Los Banos, Laguna, Philippines.

ARS (Agricultural Research Station) (1981) *Half Yearly Report*, Maha 1980/81. Maha Illuppallama, Sri Lanka, 105pp.

ARS (Agricultural Research Station) (1982) *Half Yearly Report*, Maha 1981/82. Maha Illuppallama, Sri Lanka, 86pp.

Aryeetey, A.N. (1973) Chemical weed control in rice in Ghana. *Ghana Journal of Agricultural Science* 6, 199–204.

Barker, R. and Herdt, R.W. (with B. Rose) (1985) *The Rice Economy of Asia.* Resources for the Future Inc., Washington, D.C., USA and International Rice Research Institute, Los Banos, Laguna, Philippines.

Bartlett, H.H. (1956) Fire, primitive agriculture and grazing in the tropics. In: Thomas, W.L., Jr. (ed.) *Man's Role in Changing the Face of the Earth*, University of Chicago Press, USA, pp692–702.

Bhagat, R.K., Prasad, S.C., Sinha, P.N. and Singh, A.P. (1977) Effectiveness of pre-emergence application of weedicides in upland rice. *Indian Journal of Weed Science* 9, 9–13.

Biswas, S.B. and Khan Bassi, H.P.K.V.V. (1982) Studies on herbicide–soil interaction and optimum dose of butachlor for weed control in rainfed upland rice. *Abstracts of Papers*, Annual conference of the Indian Society of Weed Science, Haryana Agricultural University, Hissar, India, p12.

Borgohain, M. and Upadhaya, L.P. (1980) Efficiency of butachlor, propanil and 2,4–D in controlling weeds in upland rice. *Indian Journal of Weed Science* 12, 145–50.

Brown, F.B. (1969) *Upland Rice in Latin America*, International Rice Commission Newsletter 18 (1), 1–5.

Camus, J.S. (1921) Rice in the Phillipines. Bulletin No. 37. Bureau of Agriculture, Department of Agriculture and Natural Resources, Manila, Philippines, 90pp.

Cancian, F. (19782) *Change and Uncertainty in a Peasant Economy.* Stanford University Press, California, USA.

Castin, E.M., Janiya, J.D., Pablico, P.P. and Moody, K. (1983) Effect of tillage on upland rice weed control. *Proceedings of the 9th Asian–Pacific Weed Science Society Conference*, Manila, Philippines, pp327–48.

Cates, A.H. (1967) A practical approach to weed control in the south west Pacific. *Proceedings of the 1st Asian–Pacific Weed Control Interchange*, Honolulu, Hawaii, USA, pp11–15.

Chavez, R.S.C. (1989) 'Cultivar competitiveness and weed control in upland rice.' M.S. thesis, University of the Philippines, Los Banos, Laguna, Philippines.

Clark, C. and Haswell, M. (1970) *The Economics of Subsistence Agriculture.* 4th edn, Macmillan, London, England.

Crafts, A.S. and Robbins, W.H. (1962) *Weed Control.* 3rd edn. McGraw-Hill, New York, USA.

Curfs, H.P.F. (1975) Soil preparation and weed control for upland and irrigated rice growing. In: *Report on the expert consultation meeting on the mechnization of rice production.* International Institute of Tropical Agriculture, Ibadan, Nigeria, pp79–86.

Dash, B. and Singh, B. (1984) Effect of weed control methods on growth and yield of unpuddled rice. *Indian Journal of Weed Science* 16, 11–16.

De Datta, S.K. (1977) Weed control in rice in southeast Asia: methods and trends. *Philippine Weed Science Bulletin*, 4, 39–63.

de Walt, B.R., de Walt, K.M., Adelski, E., Duda, S., Fordham, M. and Thompson, K. (1985) Cropping systems in Pespire, southern Honduras. *Intsormil Project Preliminary Report No. 1*, University of Kentucky, Lexington, Kentucky, USA.

Deshmukh, M.R., Trivedi, K.K. and Tiwari, J.P. (1987) Response of upland rice to weed control methods. *International Rice Research Newsletter* 12(4), 47–8.

Dobby, E.H.G. (1955) Padi landscapes of Malaya. *Malayan Journal of Tropical Geography* 6, pp94.

Domingo, I.S. and Palis, G.T. (1966) Chemical weed control experiment in upland rice at La Granja Experiment Station. *Philippine Journal of Plant Industry* 31, 145–56.

Elahi, M.N.E., Quddus, A., Naseem, S.B., Khan, A.H., Karim, S.M.R., Bhuiyan, A.M., Alam, M., Faruque, A.K.M., Alam, S. and Magor, N.P. (1985) BRRI farming systems research program. *Report of the 11th Asian Rice Farming Systems Working Group Meeting.* BRRI, Joydebpur, Bangladesh and IRRI, Los Banos, Laguna, Philippines, pp28–78.

Ellis, K.C. and Haws, L.D. (1976) Applied rice research in northeast Brazil. *International Rice Research Newsletter* 1(1), 22.

Fagade, S.O. (1976) Evaluation of chemical and mechanical weed control methods in upland rice culture. *International Rice Commission Newsletter* 25(1), 44–5.

Gassaver, E. and Fion, D. (1988) Rain-fed rice growing in Guatemala. *Agricultural News* 3/88, 20–4.

Ghosh, B.C., Singh, H.B., Sharma, H.C. and Singh, M. (1975) Effect of time of weed removal on the performance of upland rice (Ratna) under rainfed conditions of Varanasi. *Indian Journal of Weed Science* 7, 42–8.

Gogoi, A.K. (1989) The status of weed control in some of the major crops of Assam. *Abstracts of Papers*, 12th Asian-Pacific Weed Science Society Conference, Seoul, Korea, p116.

Gopalakrishna Pillai, K. and Rao, M.V. (1974) *Current status of herbicide research on rice in India.* Paper presented at the International Rice Research Conference, 22–25 April 1974, IRRI, Los Banos, Laguna, Philippines.

Gunasena, H.P.M. (1974) Current results from cooperative weed control trials in Sri Lanka. *Evaluation of herbicides for rainfed upland rice.* Paper presented at the International Rice Research Conference, 22–25 April 1974, IRRI, Los Banos, Laguna, Philippines.

Hill, R.D. (1982) Agriculture in the Malaysian region. *Geography of World Agriculture II.* Research Institute of Geography, Hungarian Academy of Science, Budapest, Hungary.

IITA (International Institute of Tropical Agriculture) (1983) *Annual Report for 1982.* Ibadan, Nigeria, 217pp.

IITA (International Institute of Tropical Agriculture) (1984) *Annual Report for 1983.* Ibadan, Nigeria, 218pp.

Indhapun, P., Schiller, J.M. and Tanomsuk, S. (1979) Weed competition and factors affecting the efficiency of herbicides in upland rice. *Research papers presented at the 17th National Conference on Agriculture and Biology*, 5–7 February 1979,

Kasetsart University, Bangkok, Thailand, pp15–28.

Janiya, J.D. and Moody, K. (1979) *The effect of time of land preparation and weed control on weeds and yields of upland rice.* Paper presented at the 10th Annual conference of the Pest Control Council of the Philippines, 2–5 May 1979, Manila, Philippines.

John, V.T. and Kalode, M.B. (1981) Cultivation practices and the impact of modern varieties. *A review of pest, disease and weed complexes in high yielding varieties in Asia and Pacific.* No. RAPA 45, FAO Regional Office, Bangkok, Thailand, pp3–39.

Juliano, J.B. (1940) Viability of some Philippine weed seeds. *Philippine Agriculture* 29, 313–26.

Kawano, K., Sanchez, P.A., Nurena, M.A. and Velez, J.R. (1972) Upland rice in the Peruvian jungle. *Rice Breeding.* IRRI, Los Banos, Languna, Philippines, pp637–43.

Kehinde, J.K. (1985) Effect of crop establishment techniques and seed rate on upland rice yield and weed growth. *International Rice Research Newsletter* 10(6), 24–5.

Kennard, C.P. (1973) Control of weeds in direct-seeded rice with some of the newer herbicides. International Rice Commission Newsletter 22(1), 15–21.

Klingman, G.C. (1961) *Weed Control as a Science.* Wiley, New York.

Kolhe, S.S. and Mittra, B.N. (1981) Weed control in direct seeded upland rice. *Proceedings of the 8th Asian–Pacific Weed Science Society Conference,* Bangalore, India, pp67–72.

Liboon, S.P., Zandstra,. H.G. and Price, E.C. (1978) C171–136 upland rice compared with Dagge in a rice-corn cropping system in the Philippines. *International Rice Research Newsletter* 3(1), 17–18.

Mallick, R.N., Shresth, R.B., Upadhyay, B.P. and Choudhary, F.M. (1974) Cooperative weed control experiments in Nepal. Paper presented at the International Rice Research Conference 22–25 April 1974, IRRI, Los Banos, Laguna, Philippines.

Mangunsoekardjo, S. and Kadnan, N. (1971) Weed control in upland rice. *Proceedings of the 1st Indonesian Weed Science Conference,* Weed Science Society of Indonesia, Contribution No. 1, Bogor, Indonesia, pp133–6.

Mayer, J. and Chabrolin, R. (1972) Faut-il rehabiliter le riz pluvial? *Technical Development* 1, 34–8.

Merlier, H. (1973) *Weed Control for Upland Rice.* Paper presented at the Seminar on Plant Protection for the Rice Crop, 21–29 May 1973, West African Rice Development Association, Monrovia, Liberia.

Merlier, H. (1983) Weed control research in dryland rice in the Ivory Coast. *Weed Control in Rice.* IRRI, Los Banos, Laguna, Philippines, pp139–44.

Mesquita, M.L.R. (1988) 'Influence of land preparation and weed control methods on weed growth and yield of upland rice.' M.S. thesis, University of the Philippines, Los Banos, Laguna, Philippines.

Modgal, S.C. and Singh, R.P. (1980) Weed control through herbicides in direct seeded rice. *Indian Journal of Weed Science* 12, 80–6.

Moody, K. (1977) *Weed control in rice.* Lecture note 30. 5th Biotrop Weed Science Training Course, 14 Nov–23 Dec 1977. Rubber Research Institute, Kuala Lumpur, Malaysia, pp374–424.

Moody, K. (1982) Changes in weed population following forest clearing in Oyo State, Nigeria. *Tropical Agriculturist* 59, 298–302.

Moody, K. (1989) *Weeds Reported in Rice in South and Southeast Asia.* IRRI, Los Banos, Laguna, Philippines, 442pp.

Mukhopadhyay, S.K. (1967) Studies on weed control in rice crop: I. Weed control in rice by Stam F34, 2,4–D and cultural methods. *Indian Agriculture* 11, 33–9.

Mukhopadhyay, S.K., Ghosh, B.C. and Maity, H. (1971) Weed problems in upland rice and approaches to solve the problem by use of new herbicides. *Oryza* 8, 269–74.

Mukhopadhyay, S.K., Ghosh, A.K. and De Datta, S.K. (1989) Weed management in upland rice by chemical methods. *Abstracts of papers*, 12th Asian–Pacific Weed Science Society Conference, Seoul, Korea, p61.

NAS (National Academy of Sciences) (1979) *Tropical Legumes*: Resources for the Future. Washington, D.C., USA.

Noguchi, K. (1989) Determination of post-seeding period of weed-free field condition in upland crops. *Proceedings of the 12th Asian–Pacific Weed Science Society Conference*, Seoul, Korea, pp207–11.

Olofintoye, J.A. (1989) Tillage and weed control practices for upland rice (*Oryza sativa* L.) on a hydromorphic soil in the Guinea savanna of Nigeria. *Tropical Agriculture* 66, 43–8.

Oyenuga, V.A. (1967) *Agriculture in Nigeria.* FAO, Rome, Italy.

Pande, H.K. and Bhan, V.M. (1964) Effect of varying degree of soil manipulation on yield of upland paddy (*Oryza sativa*) and on associated weeds. *Canadian Journal of Plant Science* 44, 376–80.

Pande, H.K. and Bhan, V.M. (1966) Effect of depth of tillage on yield of upland paddy and on associated weeds. *Experimental Agriculture* 2, 225–32.

Pande, H.K., Tripathi, R.S. and Sudhakar, N. (1983) Effective methods of weed control in upland rice. *Proceedings of the 9th Asian–Pacific Weed Science Society Conference*, Manila, Philippines, pp156–61.

Parker, C. (1968) Weed problems in India, West Pakistan and Ceylon. PANS(C) 14, 217–28.

Parry, G. (1974) *Rice Production in Ghana.* Paper presented at a Project Manager's meeting, February 1974, West African Rice Development Association, Monrovia, Liberia.

Ramamoorthy, K. and Ali, A.M. (1985) *Echinochloa colonum* (L.) Link competition and control in upland bunded rice and pulse systems. *Abstracts of papers*, 10th Asian–Pacific Weed Science Society Conference, Chiangmai, Thailand,

Renaut, G. (1972) Contribution to the study of the weeding of upland rice in Ivory Coast. *Agronomie Tropicale* 27, 221–8. (In French, English abstract)

Ruthenberg, H. (1980) *Farming Systems in the Tropics.* 3rd edn. Clarendon, Oxford, England, 424pp.

Sabio, E.A., Fisher, H.H. and Pastores, R.M. (1981) Results from retesting a production technology package based on C22 rice in Cavite. *Philippine Journal of Weed Science* 8, 41–8.

Sahai, B. and Bhan, V.M. (1982) Competition for nitrogen between weeds and drilled rice – effect of time of weed removal. *Abstracts of papers*, Annual conference of the Indian Society of Weed Science, Haryana Agricultural University, Hissar, India, p14.

Sahai, B., Bhan, V.M. and Balyan, R.S. (1983) Weed emergence and competition in drilled upland rice in India. *Tropical Pest Management* 29, 125–8.

Sanchez, P.A. and Nurena, M.A. (1970) *Upland rice improvement under shifting cultivation systems in the Amazon basin of Peru.* Agricultural Experiment Station, North Carolina State University, Raleigh, North Carolina, USA.

Sankarama, A. (1986) *Rice Production Manual Theory and Practice.* Madras, India, 126pp.

Schwerzel, P.J. (1976) The effect of depth of burial in the soil on the survival of some common Rhodesian weed seeds. *Rhodesia Agricultural Journal,* 73, 97–9.

Selvadurai, S. and Arope, A.B. (1969) *Socio-economic Study of Padi Farms in Province Wellesly, 1968.* Agronomy Branch, Division of Agriculture, Ministry of Agriculture and Cooperatives, Malaysia, 107pp.

Sengupta, K., Bhattacharyya, K.K. and Chatterjee, B.N. (1985) Intercropping of upland rice with blackgram (*Vigna mungo* (L.)). *Journal of Agricultural Science* 104, 217–21.

Shelke, D.H., Bhople, R.H. and Jadhav, N.S. (1986) Integrated weed management in upland irrigated rice in Marathwada. *International Rice Research Newsletter* 11(4), 39–40.

Simbolon, H. (1985) Competitive effect on weed species on rice in Taman Jaya, West Java. *Penelitian Pertanian* 5(2), 71–5. (In Indonesian, English abstract)

Singh, B. and Dash, B. (1984) Effects of methods of weed control on weeds and uptake of nitrogen by weeds in direct seeded unpuddled rice. *Agricultural Science Digest* 4, 1231–3.

Singh, G., Mandal, K. and Singh, B. (1983) Effect of chemical and cultural weed control methods on yield of drilled rice. *Proceedings of the 9th Asian–Pacific Weed Science Society Conference,* Manila, Philippines, pp203–9.

Singh, G., Yadav, S.R. and Singh, D. (1985) Crop–weed competition studies in upland rice. *Abstracts of Papers,* 10th Asian–Pacific Weed Science Society Conference, Chiangmai, Thailand, p13.

Singh, M. and Singh, M. (1989) Cultural-cum-herbicidal weed management studies in upland rice. *Proceedings of the 12th Asian–Pacific Weed Science Society Conference,* Seoul, Korea, pp345–51.

Singh, R. and Shrivastava, S.K. (1989) Weed control in direct seeded rice under upland conditions of Chhattisgarh, India. *International Rice Research Newsletter* 14(5), 32.

Singh, R., Singh, Y. and Singh, R.K. (1988) Performance of different tillage practices and weed control treatments on rainfed rice. *Abstracts of papers,* Biennial conference of the Indian Society of Weed Science, Assam Agricultural University, Jorhat, India, p4.

Singh, R.P. and Singh, R.K. (1983) Weed growth in upland rice as affected by weed control treatments. *Proceedings of the 9th Asian–Pacific Weed Science Society Conference,* Manila, Philippines, pp210–16.

Singh, T.N., Singh, G. and Singh, H.P. (1982) Chemical weed control in dryland rice. *International Rice Research Newsletter* 7(5), 21–2.

Singlachar, M.A. and Chandrashekar, G. (1977) Herbicides for direct-sown upland rice. *International Rice Research Newsletter* 2(2), 7.

Suprapto, H. (1980) *Intraspecific and interspecific competition of upland rice.* Southeast Asian Regional Centre for Tropical Biology, Bogor, Indonesia.

Tosh, G.C., Reddy, B.S. and Nanda, K.C. (1981) Integrated weed management in dryland rice. *International Rice Research Newsletter* 6(15), 20.

Uichanco, L.B. (1959) (ed.) *Field crops.* 2nd Revised edition of Philippine Agriculture, No. 1. College of Agriculture, University of the Philippines, Laguna, Philippines, 920pp.

Velasco, J.R., Vega, M.R., Llena, P.A. and Obien, S.R. (1961) Studies on weed control in upland rice. *The Philippine Agriculturalist* 44, 373–93.

Watters, R.F. (1958) Cultivation in Old Samoa. *Economic Geography* 34, 338–51.

Webster, C.C. and Wilson, P.N. (1966) *Agriculture in the Tropics.* Longmans, London, England.

Went, F.W., Juhren, G. and Juhren, M.C. (1952) Fire and biotic factors affecting germination. *Ecology* 33, 351–64.

Wijewardene, R. (1980) Energy-conserving systems for the humid tropics. *Agricultural Mechanization in Asia* 11, 47–53.

Wilson, G.F. and Kang, B.T. (1981) Developing stable and productive biological cropping systems for the humid tropics. In: Stonehouse, B. (ed.) *Biological Husbandry.* Butterworths, London, England, pp193–203.

Yamoah, C.F., Agboola, A.A. and Mulongoy, K. (1986) Decomposition, nitrogen release and weed control by prunings of selected alley cropping shrubs. *Agroforestry Systems* 4, 239–46.

Chapter 12
A General Overview of Kenya's Most Problematic Grassy Weeds

R.W. Michieka
Jomo Kenyatta University, College of Agriculture and Technology, P.O. Box 62000, Nairobi

Introduction

This chapter describes some important grass weeds in Kenya and the associated crops. Grasses pose major problems in Kenya's arable land. Equally important are many broadleaf weeds which are capable of smothering young crops, causing complete loss. Weeds cause more crop loss than any other pest or disease. The number of weed species in a given crop situation varies from place to place depending on various environmental factors. The extent of weed competition will depend on the pressure and abundance of the weeds present in a given situation.

Grassy weeds of arable land

The most important grassy weeds found in Kenyan arable land are listed below. Selected crops which are infested by the weeds are indicated and the extent of infestation is also given. The list is not complete but gives an indication of the current status of grassy weeds in Kenya.

Pennisetum clandestinum Chiov.

This is a problem weed in several crops including coffee, tea, maize, pyrethrum, carnations and citrus fruits. The weed establishes well in perennial crops where weeding frequency is less than in annual crops.

This perennial weed establishes itself fast during the wet season and survives dry conditions. It is a problem weed in many parts of Kenya especially at high altitudes (1400–3300 m) where rainfall is adequate. The weed is aggressive and colonizes an area in a short period of time. In some cases, *P. clandestinum* is an extremely important fodder crop in Kenya.

Rottboellia cochinchinensis (Lour) W.D. Clayton

Rottboellia infests many graminaceous crops including maize, sugarcane, sorghum and millets. It is a problem weed in cotton-growing areas of eastern and western Kenya. The weed comes late in the season and thrives best in heavier soils which have adequate moisture. This annual weed infests many hectares of the country despite the availability of techniques for controlling it. Repeated, frequent cultivation reduces the weed problem.

Avena fatua L.

This annual grass is a problem weed in the Kenyan highlands where wheat, oats and barley are grown. *Avena* spp. (wild oats) depress yields and contaminate the harvested grain.

It is estimated that over 50,000 ha are infested with wild oats including most of the wheat-growing areas. Spread of this weed has been facilitated through poor husbandry practices. Other *Avena* species which are becoming increasingly important include: *A. sterilis* (L.) ssp. *ludoviciana* (Dur.) Nyman., *A. sterilis* var. *maxima* Perez Lara. and *A. sterilis* var. *macrocarpa* Moench. There is an urgent need to prevent further invasions of *A. fatua* into new areas. It is becoming a problem weed in maize growing highlands of Kenya.

Bromus spp.

Bromus diandrus Roth. and *B. pectinatus* Thunb. have become two of the most problematic annual grass weeds in wheat and barley in Kenya. They have invaded many hectares of land and are difficult to control.

Eleusine indica (L.) Gaertn.

E. indica is a common annual grass in many crops and on many soils. It can be mistaken for finger millet (*E. coracana* (L.) Gaertn.) during the early stages of weeding. The weed is particularly problematic in millets, rice and sorghum.

Setaria verticillata (L.) P. Beauv.

This is a common weed in several annual and perennial crops including maize, wheat, beans, potatoes, sugarcane, coffee, tea and pyrethrum. In most situations, *S. verticillata* occurs in much greater densities than other grasses and broadleaved weeds. The weed survives well on all types of soils, particularly on those that are well drained.

Cynodon dactylon (L.) Pers.

A problem perennial grass, *C. dactylon* is known to many farmers as a difficult-to-control weed in many crops, especially perennials. It infests virtually all arable areas of Kenya and is particularly noted in coffee, tea, maize, potatoes, bananas, fruit orchards and pyrethrum fields.

C. dactylon is a major problem on small-scale farms where it can suppress entire fields of maize, sorghum and beans, if not controlled adequately. It is often confused with the closely related stoloniferous, non-rhizomatous *C. nlemfuensis* Vanderyst.

Digitaria abyssinica (A. Rich.) Stapf

D. abyssinica is a notorious rhizomatous weed, infesting many areas. It is particularly problematic in small-scale farms of Kenya where it infests cotton, beans, rice, maize and other annual crops. It is also a problem weed of perennial crops, including sugarcane, pineapple and coffee.

Other grass weeds

The following grasses also cause problems in the Kenyan arable lands:

Eragrostis tenuifolia (A. Rich.) Steud. – a tufted annual species.
Hyparrhenia hirta (L.) Stapf. – a widespread, variable perennial.
Brachiaria eruciformis (J.E. Smith) Griseb. – an annual species.
Echinochloa colona (L.) Link. – this annual grass is often the dominant weed of rice. It occurs in cotton and other crops especially under irrigated conditions.
Themeda triandra Forssk. – a widespread perennial species.
Eleusine jaegeri Pilg. – a tufted perennial weed of pastures in the highlands.
Oryza barthii A. Chev. – annual wild rice, grows in shallow water as a weed of rice.

Grassy weeds in Kenya are quite serious as long as the small-scale farmers practise only one weed control method. The amount of seed or perennating organs will carry on increasing as new arable lands are opened for agriculture. Control methods which combine two or more techniques would help to reduce the risks posed by grassy weeds.

Bibliography

Ivens, G.W. (1989) *East African Weeds and their Control.* 2nd edn. Oxford University Press, Nairobi, Kenya, 289pp.

Terry, P.J. and Michieka, R.W. (1987) *Common Weeds of East Africa.* Magugu ya Afrika Mashariki, FAO, Rome, 184pp.

Chapter 13
Grassy Weeds in the Philippines

J.V. Pancho

Institute of Biological Sciences, College of Arts and Sciences, University of the Philippines at Los Baños, Laguna, Philippines

Introduction

The Philippines are a group of approximately 7,100 islands and islets which are situated off the southeast coast of Asia, separating the South China Sea from the Pacific Ocean. In general, the Philippines have a maritime and tropical climate with relatively high humidity, mild temperature, abundant rainfall and gentle winds. This chapter reviews some of the more important weeds found in the area.

Weed species

Echinochloa P. Beauv.

The genus *Echinochloa* is a complex of widely and capriciously distributed weeds occurring in rice throughout the world. It embraces plants that are well adapted to wet conditions and often grow in water throughout their life cycle. About 50 species have been described (Michael, 1983). In the Philippines about seven species and two varieties are reported as common paddy field weeds in rice (Pancho and Obien, 1990).

Echinochloa colona (L.) Link

This species is well known as a weed of rice in subtropical and tropical regions throughout the world and is common in the Philippines. Propagation of this species is by seed. It is an alternate host of the root-knot nematode *Meloidogyne incognita* (Valdez, 1968), the rice green leafhoppers, *Nephotettix virescens* (Viswanathan and Kalode, 1981), and *N.*

nigropictus (Siwi, 1986), and the rice *Hoja blanca* virus. Wathanakul (1964), Mishra *et al.* (1973), and Rao and John (1974) found this species to be susceptible to tungro virus in artificial inoculation tests.

Echinochloa glabrescens Munro (*E. crus-galli auct. non* (L.) Beauv.)

Widespread from the Indian subcontinent through mainland southeast Asia, China to Taiwan, Korea and southern Japan. It has also been recorded in Togo, West Africa (Scholz, 1978). It is widespread in lowland rice fields of the Philippines. Propagation is by seed. It is an alternative host of the rice green leafhopper and the rice black bug (*Scotinophora coarctata*). IRRI (1985) found this species susceptible to tungro associated viruses.

Echinochloa crus-galli (L.) Beauv. ssp. *hispidula* (Retz.) Honda

Native of Asia now established in many tropical countries. It is a common weed in lowland rice fields of the Philippines. Propagation is by seed.

Echinochloa oryzoides (Ard.) Fritsch

This species is distinguished by its ovate spikelets and the more or less horizontally hanging inflorescence at maturity. Awns reach up to 50 mm.

According to Michael (1983) this species occurs in rice-growing areas throughout the world. It occurs in the Indian subcontinent, Sri Lanka, China, Korea, Japan, Manchuria, the Soviet Union, southeast Asia, Afghanistan, Iran, Iraq, Egypt, Argentina, Peru, the states of Louisiana and California in the USA, Australia, New Zealand and probably in all European rice-growing countries. In the Philippines, it is abundant from central to northern Luzon. It is a native of Asia (including Malaysia). Propagation is by seed.

Echinochloa crus-galli (L.) Beauv. var. *austro-japonensis* Ohwi

The close, short with more or less erect branch inflorescence and the ovate-elliptical or elliptical spikelets with usually short awns are characteristics of this species.

This species is found mainly in eastern Asia, extending south from Taiwan, the Ryukyu Islands, and the adjacent parts of mainland Asia to the higher, cooler parts of southeast Asia, Papua New Guinea and the Pacific Islands. In the Philippines, it is confined to the rice terraces and vegetable areas in the highlands of Luzon. Propagation is by seed.

Echinochloa pyramidalis (Lam.) Hitch. & Chase

This species is of African origin but was introduced to India, tropical

America and Australia. It was recently introduced into the Philippines where it is abundant in rice fields in the Laguna areas. Propagation is by seed.

Echinochloa stagnina (Retz.) Beauv.

E. stagnina is abundant in Madagascar and tropical Africa. It is also found in India, Vietnam, Thailand, Indonesia and the Philippines. It is usually associated with deep water and is of little significance in rice fields but an important weed of floating rice. Propagation is by seed.

Echinochloa picta (Koen.) Michael

This species is of Asian origin, occurring from India through southeast Asia to New Guinea and the Cape York Peninsula in Australia. It also occurs in East Africa. It is found throughout the Philippines, in marshy places and around borders of rice fields. It is abundant in the rice fields of Cagayan Province to Tarlac. Propagation is by seed.

Paspalum Linnaeus

Paspalum distichum L.

This species is widely distributed in the tropics and subtropics of the world. It is found throughout the Philippines, in rice fields, along irrigation ditches and in open, poorly-drained areas. It is a serious weed in lowland rice fields. Propagation is by seed and division of culms. It is an alternate host of the rice green leafhopper, *Nephotettix virescens* (Viswanathan and Kalode, 1981). In India, Mishra *et al.* (1973) and Rao and John (1974) found this species to be susceptible to tungro virus in artificial inoculation tests.

Ischaemum L.

Ischaemum rugosum Salisb.

This is a major weed of rice throughout the world (Holm and Herberger, 1970). In Latin America, it is reported as a common weed in wetland rice in eight countries and of dryland rice in three countries (Gonzales *et al.*, 1983). It is also one of the important weeds in rice in the Caribbean (Gonzales *et al.*, 1983) and in tropical Asia (Holm *et al.*, 1977).

I. rugosum is an alternate host of the viruses causing rice leaf gall and corn leaf gall, and of *Pyricularia* sp. It is native to tropical Asia.

***Rottboellia* L. f.**

Rottboellia cochinchinensis (Lour.) W.D. Clayton (= *R. exaltata* L. f.)

It is found in tropical Africa and Asia, Malaysia, and Australia. It occurs in pastures, cultivated fields, lightly shaded plantations, borders of tobacco plantations, and occasionally along dikes in lowland rice fields and is a prolific seed producer. Throughout the Philippines it is found in open, well-drained grasslands. The plant is host for the nematodes *Pratylenchus zeae* and *Tylenchorhynchus martini* (Anon., 1978). Propagation is by seed.

***Imperata* Cyrillo**

Imperata cylindrica (L.) Raeuschel var. *major*

I. cylindrica occurs in east and southern Africa through to Asia, Malaysia and Australia. It is found throughout the Philippines in open, dry lands from sea level to 2,300 m, often occupying large areas. It is an alternate host of the nematode *Rotylenchulus reniformis* (Anon., 1978), and the rust-fungus *Puccinia rufipes* in Thailand. The species is used in various ways in medicine (Co and Taguba, 1984). The biology, distribution and habitat, agricultural importance, commercial and other uses of this species are well documented (Holm *et al.*, 1977). Propagation is by seed and rhizome fragments. It is host of the rice tungro (RTV) – associated viruses (IRRI, 1988).

***Digitaria* Heister *ex* Fabricius**

Digitaria ciliaris (Retz.) Koel. (= *D. sanguinalis* L. Scop.)

This species is found extensively in the tropics and subtropics. In the Philippines, it is a very common weed in waste places, cultivated areas and along dikes and canals in paddy fields, from sea level to 2,400 m. Propagation is by seed. It is host of the rice tungro (RTV) – associated viruses (IRRI, 1988).

***Leersia* Swartz**

Leersia hexandra Sw.

This species is pantropic and is found throughout southeast Asia, in open fields with poorly drained soil and along dikes and open areas in lowland rice fields and is locally abundant. It propagates by division of rootstocks and seeds. It is an alternate host of the rice green leafhoppers, *Nephotettix nigropictus* (Kalode, 1983), and *N. malayanus* (Inoue, 1986). Rivera *et al.* (1969) found this species to be susceptible to tungro virus in artificial inoculation tests. It is also a secondary host of the fungus *Pyricularia oryzae* Cav. (IRRI, 1988).

Brachiaria (Trinius) Grisebach

Brachiaria mutica (Forssk.) Stapf (= *Panicum purpurascens* Raddi)

This species is found in tropical Africa and America and was introduced in most tropical countries as a fodder grass. It is often a serious paddy rice field weed in some parts of the Philippines. Propagation is by seed. It is a secondary host of the rice blast fungus *Pyricularia oryzae* Cav. and rice tungro (RTV) – associated viruses (IRRI, 1988).

References

Anon., (1978) Survey, identification and host parasite relationships of plant parasitic nematodes associated with major economic crops in the Philippines. *Technical Report* NSDB-Project No. 7314 Ag (2-275-53), (Unpublished).

Co, L.L. and Taguba, Y.B. (1984) Common medicinal plants of the Cordillera region. *Services and Training in the Cordillera Region.* Community Health Education, Baguio City, 290pp.

Gonzales, J., Garcia, E. and Perdomo, M. (1983) Important rice weeds in Latin America. *Weed Control in Rice*, IRRI, Los Banos, Laguna, Philippines, pp19–32.

Holm, L. and Herberger, J. (1970) Weeds of tropical crops. *Proceedings of the 10th British Weed Control Conference*, Brighton, England, pp132–49.

Holm, G., Plucknett, D.L., Pancho, J.V. and Herberger, J.P. (1977) *The World's Worst Weeds: Distribution and Biology.* The University Press of Hawaii, Honolulu, 609pp.

Inoue, H. (1986) Biosystematic study on the genus *Nephotettix* occurring in Asia. *Bulletin of Kyushu Agricultural Experiment Station*, 24, 149–237.

IRRI (International Rice Research Institute) (1985) *Annual Report.* P.O. Box 933, Manila, Philippines, 554pp.

IRRI (International Rice Research Institute) (1988) *International Rice Research Newsletter* 13(4), 30–33, 37.

Kalode, M.B. (1983) Leafhopper and planthopper pests of rice in India. *Proceedings of 1st International Workshop on Leafhoppers and Planthoppers of Economic Importance.* pp25–45.

Michael, P.W. (1983) Taxonomy and distribution of *Echinochloa* species with special reference to their occurrence as weeds. *Weed Control in Rice.* IRRI, Los Banos, Laguna, Philippines, 291–326.

Mishra, M.D., Ghosh, A., Nazi, F.R., Basu, A.N. and Raychaudhuri, S.P. (1973) The role of graminaceous weeds in the perpetuation of rice tungro virus. *Journal*

of the Indian Botanical Society 52, 176–83.
Pancho, J.V. and Obien, S.R. (1991) *Manual of Paddy Field Weeds in the Philippines.* (In Press)
Rao, R.D., V.J.P. and John, V.T. (1974) Alternate hosts of rice tungro virus and its vectors. *Plant Disease Reporter* 58, 856–60.
Rivera, C.T., Ling, K.C. and Ou, S.H. (1969) Suscept range of rice tungro virus. *Philippine Phytopathology,* 5. 16–17. (Abstract)
Scholz, H. (1978) Zwei Gramineen neu für Togo: *Echinochloa glabrescens* und *Panicum laxum. Willdenowia* 8, 375–7.
Siwi, S.S. (1986) Variation in morphological characteristics of *Nephotettix virescens* (Distant) (Hom., Euscelidae) from different islands in Indonesia. *Contributions,* Central Research Institute for Food Crops, Bogor, No. 75, 26pp.
Valdez, R.B. (1968) Survey, identification and host-parasite relationships of root-knot nematodes occurring in some parts of the Philippines. *The Philippine Agriculturalist* 51, 802–24.
Viswanathan, M. and Kalode, K. (1981) Studies on varietal resistance and host specificity of rice green leafhoppers. *International Rice Research Newsletter,* 6, 7–8.
Wathanakul, L. (1964) 'A study on the host range of tungro and orange leaf viruses of rice.' M.S. Thesis, University of the Philippines, Los Banos, Philippines, 35pp.

Chapter 14
Weeds of Tropical Australia

R.H. Groves

CSIRO Division of Plant Industry, GPO Box 1600, Canberra, ACT2601, Australia

Introduction

Most of the weeds in Australia have been introduced from elsewhere. Few major agricultural weeds are native, especially in the tropical region, except for the tall shrub *Acacia harpophylla* (brigalow) in southern subtropical Queensland. This brief regional overview considers mainly those major introduced weeds occurring north of the Tropic of Capricorn in the Australian states of Queensland, Northern Territory and Western Australia (Fig. 14.1). These weeds occur in cropping lands in the areas of higher rainfall near the Queensland coast or in irrigated areas of inland Queensland or Western Australia. Alternatively, they may occur in the pasturelands and rangelands inland of, and with a lower rainfall than, the cropping lands. Some weeds have been discovered only relatively recently in northern Australia and their distribution is still localized. For instance, the African plant *Cyperus aromaticus* (Navua sedge) was found in the Cairns area in 1979. The plant also occurs on several Pacific islands, including Fiji, where it is a serious weed. The potential of *C. aromaticus* to spread from Cairns is unknown as yet, however. Some major weeds of tropical regions elsewhere are not yet known to occur in the Australian tropics. This chapter concludes by briefly mentioning some grass species in this category, which have potential to become weedy should they ever reach Australia.

The review that follows emphasizes the role of grassy weeds in affecting tropical Australian ecosystems, although in some cases little is known about their biology relative to that of some other non-grassy weeds of the region.

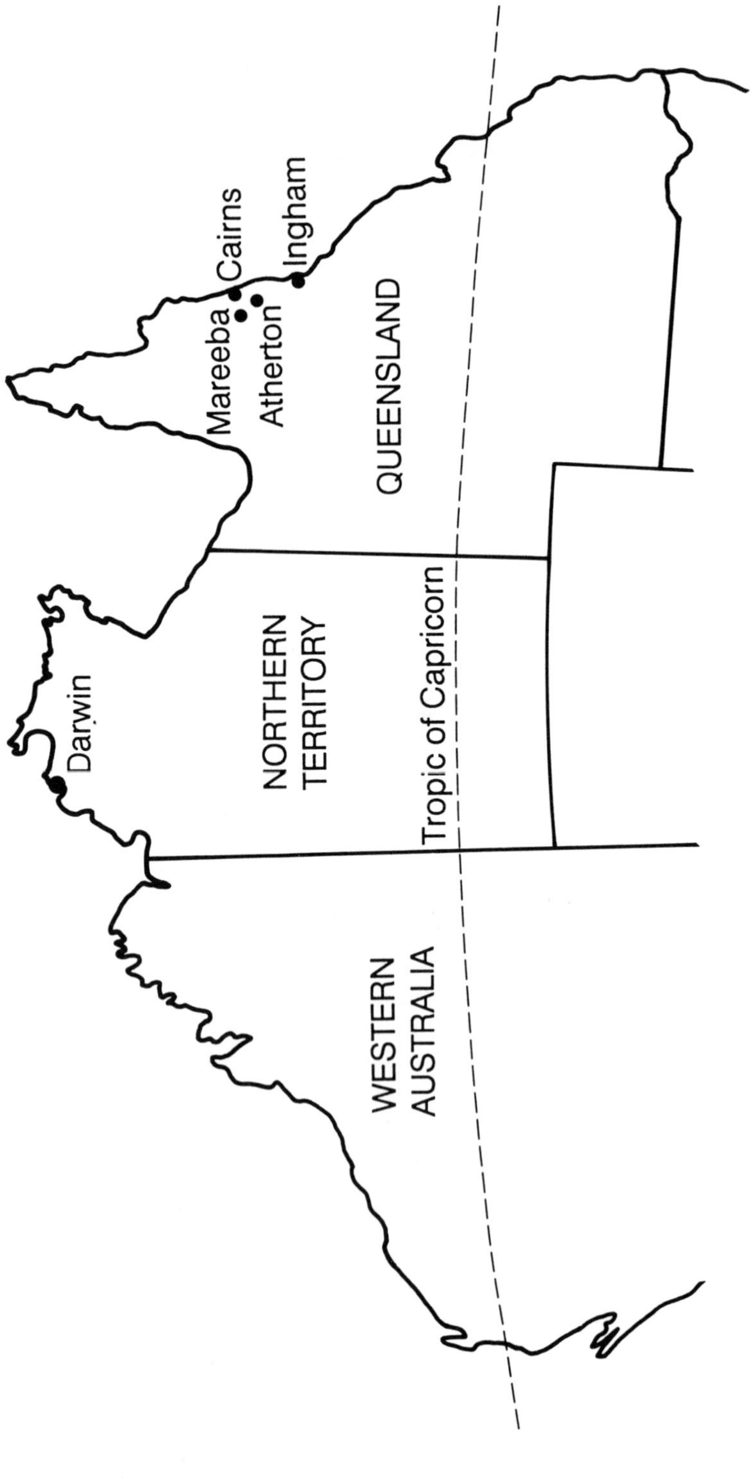

Fig. 14.1 Location of places in tropical Australia mentioned in the text.

Weeds of crops

Sugarcane is cropped on fertile soils that extend in a discontinuous narrow strip along the coastal lowlands from northern New South Wales to north of Cairns, Queensland (Fig. 14.1). Rainfall is summer-incident and usually occurs in adequate amounts for the growth of sugarcane; where it is not sufficient, the crop may be irrigated. Two weeds of sugarcane are *Rottboellia cochinchinensis* (itch-grass) in the north of this region and the legume *Mimosa invisa*, one of the sensitive plants. The former species is well known as a weed of sugarcane and other crops in places such as the Philippines (Holm *et al.*, 1977), but little seems to be known about its biology as a weed of sugarcane in Queensland as it was only identified for this area relatively recently. *M. invisa* is a serious weed of pastures as well as of sugarcane in the region to the north of Ingham (Fig. 14.1). The plant is a vine that grows rapidly, has spiny leaves and stems and can produce large numbers of small hard seeds in prickly pods. Chemical control of *M. invisa* is effective but costly. A biological control programme has commenced recently that involves release of an insect (*Scammius* sp.) and a fungus (*Corynespora cassiicola*) to reduce seed production. *Brachiaria mutica* (Para grass), grown widely as a fodder grass, may also be weedy in sugarcane crops.

The area planted to tea in northern Queensland is relatively small and located inland of the sugarcane region and at a higher altitude. There have been some new plantings recently. A wide range of weeds occurs in the early stages of plantation; these weeds are usually controlled by cultivation initially or by herbicide application. *Eleusine indica* (crowsfoot grass) and *Setaria sphacelata* are the main grassy weeds on newly planted tea estates in the region. Weeds of recently established coffee plantations in the Mareeba area (Fig. 14.1) are presumably of similar composition.

Crops such as maize, sorghum and groundnuts are widely grown, often in rotation, in coastal Queensland and northern New South Wales under rainfed or irrigated conditions. Many weeds commonly occur in these crops, of which the major species are broadleaved cruciferous weeds and, in the Atherton area, *Nicandra physalodes* (Apple of Peru). Most of these weeds are controlled adequately by cultivation and by applications of several herbicides. Grassy weeds include the annuals *Cenchrus echinatus* (Mossman River grass), *Sorghum halepense* (Johnson grass) and *S. verticilliflorum* (wild sorghum). Weeds in horticultural crops are similar and as tobacco continues to decline in area and significance in northern Queensland, horticultural crops, such as tropical fruits, cut flowers and tomatoes, are planted more widely to supply southern markets. Fruit crops are of increasing importance around Darwin, especially for export to south-east Asia.

Pennisetum pedicellatum is an annual weed of grain sorghum crops in the Northern Territory and is widespread over large areas of northern

Australia (Mott, 1980). No carryover of viable seed in the soil at the end of the wet season was found, the population in any year depending on viable seed set in the preceding year only. In crops, hygiene combined with minimum tillage and application of herbicide to kill seedlings that may germinate from seed remaining on the undisturbed soil surface gave effective control. In pasture, heavy grazing to prevent seed set led to good control of *P. pedicellatum* in the succeeding year.

In Australia, rice is mainly grown in inland southern areas, such as the Murrumbidgee Irrigation Area near Griffith, New South Wales. But for about the last 15 years, rice has also been grown for sale on the domestic market in northern Queensland. The main weeds of the crop around Mareeba are *Echinochloa colona* (awnless barnyard grass) and *Cyperus iria*, which are controlled by a post-emergent aerial application of propanil. A variety of *Oryza sativa* (so-called straw-hulled red rice) with dormant seeds is a serious contaminant of the rice crop in the Mareeba area, having been originally introduced from the United States in contaminated seed.

Cotton is grown in tropical and subtropical Queensland although the present main cotton-producing area is in northern inland New South Wales. An earlier attempt to establish a cotton-growing industry in northwest Western Australia failed, primarily for economic reasons. In all areas, the major weeds of cotton are *Echinochloa* spp. (barnyard grasses) and several species of *Xanthium* (Noogoora and Bathurst burrs). A potential change in control of *Xanthium* is the current development of a mycoherbicide for *X. spinosum* (Bathurst burr). A closely related strain of the fungus (namely *Colletotrichum* sp. aff. *xanthii*) may be sufficiently pathogenic on the Noogoora burr species complex (*X. occidentale* and *X. italicum*) to allow its future development as a mycoherbicide. Another current development in weed control in cotton in Australia is research on the transfer of genes for herbicide resistance into cotton plants with the long-term aim of reducing the impact of broadleaved weeds on cotton production. Both in cotton and rice crops in New South Wales a further recent development has been the emergence of several native plants as weeds – a species of the native herb genus *Haloragis* is resistant to the herbicides used currently in cotton and is becoming weedy, whilst in the southern rice areas, the native grass *Diplachne fusca* is spreading and becoming weedy because it is preadapted to rice cultivation and, specifically, to shallow ponding of water (McIntyre *et al.*, 1989).

In summary, the weeds of northern Australian crops are all introduced and are broadly similar to those of crops in other tropical regions. One feature that may become more apparent with time is the recent emergence of indigenous species as weeds of irrigated crops such as cotton and rice. Such species will either be resistant to the herbicides currently in use and/or be pre-adapted to the cultivation or moisture regimes pertaining to the growth of irrigated crop plants.

Weeds of pastures

The area of land used as pastures and rangelands in tropical Australia is very much larger than the relatively small pockets of cropping land situated along the coast or in inland river valleys. In addition, the economic inputs to pasture land tend to be lower. The combination of factors is reflected in the lower overall emphasis on short-term chemical control and a greater emphasis on longer-term biological control programmes. The biology of many of the candidate weeds for biological control is well known compared with the group of crop weeds. Because the former group does not include grasses as yet, most of the biological information available currently is for non-grassy weeds.

As elsewhere, *Imperata cylindrica* (blady grass) is a major problem plant on infertile land that is burned regularly. Although its distribution extends the length of coastal southeastern Australia it is really only weedy in degraded pastures in Queensland and northern New South Wales.

Of the grass weeds of tropical pastures, three are locally significant. *Themeda quadrivalvis* (grader grass), a tufted robust annual, native to India, is widespread in coastal Queensland, where it was introduced in 1935 (Miller *et al.*, 1982); it has been introduced subsequently to the Northern Territory as a contaminant of pasture seed, and also to coastal northern New South Wales. It is primarily a weed of overgrazed pastures or disturbed areas such as roadsides where it may grow up to 2 m tall – hence its common name. Mature plants are unpalatable to grazing animals. In Queensland *T. quadrivalvis* is also becoming a problem in sugarcane and in citrus orchards. Control of this grass is achieved by a combination of methods – application of paraquat, avoidance of overgrazing, low slashing and burning of pastures, prevention of seeding by high slashing at flowering, together with use of clean imported seed in resowing of pastures (Miller *et al.*, 1982). In this and other respects, *T. quadrivalvis* resembles the introduced perennial grass *Hyparrhenia rufa* (thatch grass) native to tropical Africa and also spreading currently along roadsides in Queensland, as also is *Rhynchelytrum repens* (Natal grass).

Pennisetum polystachion (mission grass) is a perennial weed of roadsides in the Darwin area. Its presence may be changing the established burning regime for the area because it does not burn until later in the season than the native annual grass *Sorghum intrans* which it is gradually replacing. Later fires may cause more damage to indigenous tree and shrub species and hence the natural environment may change. The same grass is used as a pasture species in Fiji.

The economically significant weeds of pastures in northern Australia are usually woody dicotyledonous plants, however. They include *Mimosa pigra* in the Northern Territory, *Hyptis* and *Sida* species in both Northern Territory and Queensland and, in Queensland, *Acacia nilotica* ssp. *indica* (prickly acacia), *Parthenium hysterophorus* (parthenium weed), *Cryptostegia grandiflora* (rubber vine) and *Eriocereus martinii* (harrisia cactus). All these weeds are currently candidates for biological control, with the control of *E. martinii* being the most successful to date, partly because it was one of the first for which natural enemies were imported and released. *Parkinsonia aculeata* is also a woody weed of rangelands occurring much more commonly in northwest Western Australia and Northern Territory than in Queensland. Chemical control of all these plants is also being researched currently and constitutes an essential part of an overall control programme for this major group of introduced woody weeds. In the absence of effective control for some of them at the moment, some are still spreading and undoubtedly have the potential to spread much further than their present limited distribution in Australia, as in the cases of *Mimosa pigra* (Lonsdale *et al.*, 1989) and *Parthenium hysterophorus* (Williams and Groves, 1980), to cite just two instances.

Grassy weeds of potential to tropical Australia

Michael (pers.comm.) has recently identified 23 plant species that are weeds in countries to the north of Australia but not yet known to occur in tropical Australia. Of these 23 species, eight are grasses (Table 14.1). Most are perennial weeds of flooded rice in southeast Asia. Two grasses, *Leptochloa panicea* and *L. chinensis*, arc listed by Michael as constituting the greatest potential threat to northern Australia. *Echinochloa glabrescens* is also considered a high risk because it is so abundant in the Philippines where it is sometimes the most important species affecting rice yields; it can

Table 14.1. Some grass weeds of potential quarantine concern to tropical Australia (Michael, pers.comm.).

Species	*Annual/ perennial*	*Aquatic and/ or rice weed*	*Region of origin*
Coix aquatica	p	+	southeast Asia
Digitaria fuscescens	p	–	Asia and Africa
Echinochloa glabrescens	a	+	eastern Asia
E. stagnina	p	+	Africa
Eriochloa polystachya	p	+	America
Leptochloa chinensis	a/p	+	southeast Asia
L. panicea	a	+	Asia and Africa
Sacciolepis interrupta	p	+	Asia and Africa

be easily transported in poorly-cleaned grain. Michael's list of these grass weeds is extremely important and timely in focusing the attention of quarantine authorities, research taxonomists and crop agronomists.

Concluding discussion

A large number of grasses from other regions of the tropics are naturalized in tropical Australia where they may be weedy to varying extents. Grasses introduced from tropical East Africa have the highest numerical representation (Simon, 1988), which presumably reflects climatic similarities between the two regions. The biology of some of these grasses is relatively well known, as in the cases of *Imperata cylindrica*, *Echinochloa* spp. and *Sorghum halepense*, although the information available is rarely Australian in origin. Biologically, little seems to be known about some other tropical grass weeds, such as *Themeda quadrivalvis* and *Pennisetum polystachion*, and the extent of their impact on Australian agro-ecosystems. The potential for some other grasses to become weeds in northern Australia should they ever be introduced is great, especially in the cases of several species of the genera *Echinochloa* and *Leptochloa*, which are major weeds of agriculture elsewhere in the tropics. Clearly, present information on grassy weeds is as inadequate for tropical Australia as elsewhere in the tropics.

Whilst knowledge of the biology of Australia's tropical grasses as weeds of cropped areas is inadequate, relatively more is known about some dicotyledonous weeds of pastures, as this brief review has attempted to highlight. The biology of weeds that are candidates for biological control has received considerable research attention both in Australia, where they have been introduced, and, to a lesser extent, in their native regions. Hitherto grasses have not been actively considered as candidates for biological control; perhaps the increasing attention being paid them in this regard (Evans, Chapter 3, this volume; Wapshere, 1990) will restore a more appropriate balance in research between grassy and other non-grassy weeds in tropical Australia.

Acknowledgements

Dr Peter Michael kindly made available his review paper on weeds of concern to northern Australia, prior to its publication, for which I thank him. I also thank Drs John Mott (CSIRO Division of Tropical Crops & Pastures, Brisbane) and Stella Humphries (CSIRO Division of Wildlife & Ecology, Canberra) for comments on a draft of this chapter.

References

Bishop, H.G. (1981) Grader grass – a nuisance weed. *Queensland Agriculture Journal* 107, 235–9.

Holm, L.G., Plucknett, D.L., Pancho, J.V. and Herberger, J.P. (1977) *The World's Worst Weeds. Distribution and Biology.* The University Press of Hawaii, Honolulu, 609pp.

Lonsdale, W.M., Miller, I.L. and Forno, I.W. (1989) The biology of Australian weeds. 20. *Mimosa pigra. Plant Protection Quarterly* 4, 119–31.

McIntyre, S., Mitchell, D.S. and Ladiges, P.Y. (1989) Germination and seedling emergence in *Diplachne fusca*: a semi-aquatic weed of rice fields. *Journal of Applied Ecology* 26, 551–62.

Miller, I.L., Harrison, P.G. and Rankin, M.O. (1982) Grader grass. A noxious weed. *Northern Territory Department of Primary Production Agnote Number 82/55.*

Mott, J.J. (1980) Germination and establishment of the weeds *Sida acuta* and *Pennisetum pedicellatum* in the Northern Territory. *Australian Journal of Experimental Agriculture and Animal Husbandy* 20, 463–9.

Simon, B.K. (1988) The biogeography of tropical Australian grasses. *Proceedings of the Ecological Society of Australia* 15, 267–9.

Wapshere, A.J. (1990) Biological control of grass weeds in Australia: an appraisal. *Plant Protection Quarterly* 5, 62–75.

Williams, J.D. and Groves, R.H. (1980) The influence of temperature and photoperiod on growth and development of *Parthenium hysterophorus* L. *Weed Research* 20, 47–52.

Recommendations

The participants agreed unanimously that the practice that could have the greatest impact on increasing crop yields is effective weed management. It was agreed that integrated pest management provided the most economic method for maintaining pests at acceptable levels. A co-ordinated international effort with the participation of specialists from many disciplines and from both developing and industrialized countries should be mounted to enhance our understanding of:

(i) the biology of undesirable plant species;
(ii) the positive and negative aspects of grassy weeds in agricultural production systems;
(iii) the natural enemies of grassy weeds;
(iv) the effects of natural and man-made toxins thereon.

In such an international effort, adequate resources should be allocated to the elucidation of the taxonomy of certain grassy weeds (*Echinochloa* spp. and *Oryza* spp. were cited as two such examples) because inappropriate identification causes misunderstandings when exchanging results of studies. Attention was drawn to the need to improve the exchange and use of information about grassy weeds. It was suggested that a network of active and interested research workers be established in relation to grassy weeds, similar to that for the parasitic weed *Striga*.

In order to highlight the losses that occur in food crop production systems because of grassy weeds, it was suggested that precise information on losses for various crops and regions should be collected and published.

1. It was recommended that studies for the control and management of grassy weeds should include the role of weeds as alternative hosts for pests and their predators.

2. It was agreed that the use of herbicides is likely to remain an important component of integrated weed management. It is therefore desirable to develop compounds with a minimum risk to the environment and to apply them sensibly. To maintain the needed flexibility in chemical weed control, particularly in order to avoid the development of herbicide resistance, the number of herbicides with different mechanisms of action should not be reduced. With respect to chemical herbicides it was suggested that all instructions for use should include information about antidotes and should advise against using doses above those recommended by the manufacturer.

3. The need to encourage the development of mycoherbicides was stressed.

4. Studies of the biology of grassy weeds should include dormancy and germination, and the partitioning of resources between seeds and vegetative organs in perennial grassy weeds.

5. The role of importation and movement of weed seeds and propagules as a factor in the spread of weeds and the importance of seed purity and plant quarantine, especially in developing countries, was stressed.

6. Attention was drawn to the problems of weeds in individual crops, such as upland rice and cotton, and of the roles of fallows and crop rotations in integrated weed management.

7. There was unanimous agreement that a much greater effort should be devoted to training and management and to the diffusion of information on weed control, especially in developing countries.

8. It was suggested that a small number of the most important grassy weeds, for example *Rottboellia*, should be selected as targets for increased research effort.

Index

Numbers in italic indicate illustrations. Bold indicates main subject entry.